THE BEASTS THAT HIDE FROM MAN

THE BEASTS THAT HIDE FROM MAN

SEEKING THE WORLD'S LAST UNDISCOVERED ANIMALS

Karl P.N. Shuker, Ph.D.

PARAVIEW PRESS

New York

The chapters constituting this book are based upon, adapted from, or inspired by the following articles:

SHUKER, Karl P.N. (1995). In the spotlight: the dobhar-chú. *Strange Magazine*, No. 16 (fall): 32-3.

SHUKER, Karl P.N. (1997). Meet Mongolia's death worm: the shock of the new. *Fortean Studies*, 4: 190-218.

SHUKER, Karl P.N. (1998). On the trail of a terror bird. *Wild About Animals*, 10 (November-December): 40-1.

SHUKER, Karl P.N. (1998). Stranger than fiction. *All About Cats*, 5 (September-October): 40-1.

SHUKER, Karl P.N. (1994). A belfry of crypto-bats. *Fortean Studies*, 1: 235-45.

SHUKER, Karl P.N. (1993). More mystery plants of prey. *Strange Magazine*, No. 12 (fall-winter): 9, 49, 51.

SHUKER, Karl P.N. (1997). Arboles devoradores de hombres. *Enigmas*, 3 (March): 8-16.

SHUKER, Karl P.N. (1996). Beware, beware, the fish with hair! *Wild About Animals*, 8 (June): 9.

SHUKER, Karl P.N. (1993). Hoofed mystery animals and other crypto-ungulates. Part III. *Strange Magazine*, No. 11 (spring-summer): 25-7, 48-50.

SHUKER, Karl P.N. (1996). Super-emus and spinifex men — how tall are the tales of giant birds? *Fortean Studies*, 3: 189-210.

SHUKER, Karl P.N. (1999). Sarajevo's jumping snake. *Fortean Times*, No. 123 (June): 46.

SHUKER, Karl P.N. (1999). On the trail of the tatzelworm. *Wild About Animals*, 11 (January-February): 40-1.

SHUKER, Karl P.N. (2001). Loco for a tzuchinoko. *Fortean Times*, No. 142 (February): 45.

SHUKER, Karl P.N. (1996). Never bother a beithir! *Wild About Animals*, 8 (July): 9.

SHUKER, Karl P.N. (1996). A real pen and ink. *Fortean Times*, No. 90 (September): 44.

SHUKER, Karl P.N. (1997). Chinese ink monkey...or Chinese whispers? *Strange Magazine*, No. 18 (summer): 52.

SHUKER, Karl P.N. (2001). Living unicorns, born-again basilisks, and latter-day dragons. Unsubmitted article.

SHUKER, Karl P.N. (1997). The little howling god. *All About Dogs*, 2 (July-August): 46-7.

SHUKER, Karl P.N. (1997). A new dawn for the sun dog. *All About Dogs*, 2 (May-June): 44-5.

SHUKER, Karl P.N. (1998). Cynocephali and other dog-headed denizens of mythology. *All About Dogs*, 3 (September-October): 44-5.

SHUKER, Karl P.N. (1997). Don't lose your head — it's the waheela. *All About Dogs*, 2 (March-April): 54-5.

SHUKER, Karl P.N. (1996). Monkey business. *All About Dogs*, 1 (March-April): 74-5.

SHUKER, Karl P.N. (1997). Lemurs of the lost...and found? *Wild About Animals*, 9 (March-April): 7.

SHUKER, Karl P.N. (2000). Catch a kilopilopitsofy! *Fortean Times*, No. 131 (February): 48.

SHUKER, Karl P.N. (1995). "Bring me the head of the sea serpent!" *Strange Magazine*, No. 15 (spring): 12-17.

SHUKER, Karl P.N. (1998). A supplement to Dr Bernard Heuvelmans' checklist of cryptozoological animals. *Fortean Studies*, 5: 208-29.

Cover illustration: artistic representation of the Mongolian death worm by Ivan Mackerle

Cover design by smythtype

ISBN: 1931044643

Library of Congress Catalog Number: 2003108039

TO MY AGENT, Mandy Little, of Watson, Little Ltd., London,
with my very grateful thanks for overseeing so diligently
and enthusiastically my writing career during the past ten years —
and, I sincerely hope, for many years to come.

CONTENTS

The Animals of Tomorrow?

What is the world's most cunning animal?
That which no man has ever seen.

TRADITIONAL AFRICAN PROVERB

MORE THAN 300 MAJOR NEW SPECIES OF ANIMALS WERE scientifically unveiled during the 1900s. As documented comprehensively in my book *The New Zoo: New and Rediscovered Animals of the 20th Century* (2002), they included such remarkable creatures as that famous forest-dwelling, short-necked giraffe known as the okapi (discovered in 1901); the gigantic but gentle mountain gorilla (1902) and the equally sizeable giant forest hog (1904); the extraordinary lobe-finned coelacanth (1938), belonging to an ancient lineage of fishes formerly believed to have died out at least 65 million years ago; large, spectacular birds such as the Congo peacock (1936) and Amsterdam albatross (1978); the stupendous megamouth shark (1976); a pig-like Ice Age relic called the Chacoan peccary (1974); eight different species of beaked whale (between 1908 and 1996); the formidable Komodo dragon (1912), currently the world's largest known lizard; the Cambodian wild ox or kouprey (1937); the Queen of Sheba's gazelle (1985); the buffalo-sized Vu Quang ox (1992), the largest new land mammal to be discovered for more than 50 years; the giant muntjac deer (1994); the bondegezou tree kangaroo (1994); and more than a dozen new South American monkeys (1990s and 2000s).

Moreover, judging from a vast array of well-documented, ostensibly reliable eyewitness accounts and other evidence on record, there may also be a sizeable number of spectacular animals still awaiting formal zoological detection and description in the 21st century, steadfastly remaining hidden from science despite apparently being familiar to their local human neighbors.

The official scientific investigation of these highly elusive creatures is known as cryptozoology (literally "the study of hidden animals"), a once-controversial fledgling zoological discipline that has gained increasing mainstream acceptance in recent years due to the startling number of unexpected zoological discoveries made by diligent field investigators, notably in Vietnam and South America.

Over the years, many books dealing with cryptozoological animals have been published. However, the majority of these have concentrated upon the same few, familiar examples—the Loch Ness monster and other freshwater mystery creatures, man-beasts (yeti, Bigfoot, etc.), sea monsters, and unidentified big cats in Britain and elsewhere. Conversely, throughout my own cryptozoological writings and research I have always preferred to pursue and highlight the little-realized plethora of lesser-known or decidedly obscure mystery beasts also on record—creatures that have often received only the briefest of mentions in the literature, and even then only in specialized, scarcely read, or largely forgotten journals, travelogues, historical accounts, and other esoteric sources.

Writing as I do for a wide range of publications, it has been regrettable but inevitable that many of my cryptozoological articles have been featured in English journals and magazines not available in the U.S., or in American ones not very accessible in Britain at the time of their appearance. Now, however, in response to popular demand from my worldwide readership, I have compiled in book form a selection of some of what I consider to be my most memorable articles from these disparate publications. In many cases, moreover, I have expanded and updated them to include additional information and illustrations that were not available at the time of my articles' original appearances. And continuing my long tradition for making my research sources easily accessible to my readers, there is also a comprehensive bibliography of references consulted by me during my preparation of this book that can be utilized by readers wishing to pursue further any of the subjects documented therein.

Here at last, therefore, with no geographical barriers to separate them from my international reading public, are such hitherto-abstruse creatures of cryptozoology as the Mongolian death worm and the Sarajevo jumping snake, the mimick dog and the ink monkey, the hairy lizards of New Guinea and the lost sirens of St. Helena, as well as Aztec sun dogs, Javan crypto-bats, Irish master otters, living unicorns, triangular serpents, giant eggs, dwarf emus, horned sea serpents, and a rich diversity of mystery beasts not even listed in Dr. Bernard Heuvelmans' cryptozoological checklist. In other words, allow me to unfurl before your very eyes the fauna of the future—the beasts that (still) hide from man.

CHAPTER 1

The *Dobhar-Chú*—Ireland's Mysterious Master Otter

By Glenade Lake tradition tells,
two hundred years ago
A thrilling scene enacted was,
to which as years unflow
Old men and women still relate,
and while relating dread,
Some demon of its kind may yet
be found within its bed.

ANONYMOUS, C. EARLY 1900S?—"THE DOBHAR-CHÚ OF GLENADE"

THE *DOBHAR-CHÚ*, A SUPPOSEDLY MYTHICAL BEAST from Ireland, is also called the *dobarcu*, master otter, and king otter, and was classed by English folklorist Dr. Katharine Briggs as a prototype animal representing all of its kind. In *The Anatomy of Puck* (1959), Briggs termed it the master otter, and it was evidently larger than normal otters because she stated that it was said to have appeared once at Dhu-Hill, with "about a hundred common-sized otters" in attendance. According to legend, an inch of the master otter's pelt will prevent a ship from being wrecked, a horse from injury, and a man from being wounded by gunshot or other means.

In *Myth, Legend and Romance: An Encyclopaedia of the Irish Folk Tradition* (1990), Dr. Dáithí ó hÓgáin described it as a large male otter called the king otter, reiterating much of the information presented by Briggs but also noting that it was totally white in color except for its black ear tips and a black cross upon its back, and that it never slept. Yielding an unexpected parallel with the werewolf legend, this creature could only be killed with a silver bullet, and its killer would himself die no longer than 24 hours afterwards.

DOBHAR-CHÚ DOCUMENTS

Until quite recently, the above was all that I knew concerning the *dobhar-chú*—but in 1996, one of my correspondents, librarian Richard Muirhead, then of Wiltshire, southern England, kindly supplied me with several additional sources of information. These offer a much more detailed, and sinister, insight into Ireland's most mystifying mammal.

Take, for example, the following letter, written by Miss L.A. Walkington and published by the journal of the Royal Society of Antiquaries of Ireland in 1896:

> When on a recent visit to Bundoran [in County Leitrim, northwestern Ireland], we heard a legend concerning a tombstone in the graveyard of Caldwell [Conwall], which induced us to visit the place. The story is as follows: A young married woman went to wash her clothes in a stream near the house, and an animal called by the natives a dhuraghoo (that is spelled as pronounced, but I have never seen the word written), came out of the river and attacked her. Her husband (or brother according to some accounts) missing her went to look for her, and found her dead and the beast sucking her blood. The dhuraghoo attacked the horse; for the husband seems to have been on horseback. The horse being frightened, ran away, but became exhausted at a village called from this circumstance Garronard (*garron*, a bad horse; ard, a high place). The dhuraghoo is said to have gone "through" the horse and to have killed it. It was

then speared by the husband who at the same time killed its young one. The dhuraghoo is said by some to have been an animal half wolf-dog, half-fish, by others an enormous sea-otter…Two other tombstones are shown in connexion with the story, one bearing an image of the horse, and said to be that of the husband. Perhaps some antiquary may be able to throw light on the legend and on the nature of the dhuraghoo.

In a later issue of this journal for 1896, Miss Walkington's letter drew the following response from H. Chichester Hart:

I have heard at Ballyshannon, a few miles from Bundoran, the following account of the "Dorraghow," as it was pronounced in that district. He was "The King of all the Lakes, and Father of all the Otters. He can run his muzzle through the rocks. He was as big as five or six otters." My informant thought he was long dead.

The master otter also appeared in the poem "The Old House," within a 1950s anthology, *Further Poems,* by Leitrim poetess Katherine A. Fox. The relevant lines read:

The story told of the *dobhar-chu*
That out from Glenade lake
Had come one morning years ago
A woman's life to take.

During his research, Richard Muirhead also uncovered a much longer poem, of unknown source, titled "The Dobhar-chú of Glenade" (whose first verse opened this chapter), which is devoted entirely to the master otter's deadly attack upon the hapless maiden and its fatal encounter with her vengeful husband. Regrettably, its style is somewhat lurid and turgid, as witnessed by the following excerpt:

> She having gone to bathe it seems within its
>   waters clear
> And not returning when she might her husband
>   fraught with fear
> Hastening to where he her might find when oh,
>   to his surprise.
> Her mangled form still bleeding warm lay
>   stretched before his eyes.
>
> Upon her bosom snow white once but now
>   besmeared with gore
> The Dobarcu reposing was his surfeitting been o'er.
> Her blood and entrails all around tinged with a
>   reddish hue.
> "Oh God", he cried, "tis hard to bear but what am
>   I to do".

Shakespeare it ain't, that's for sure! Nevertheless, its 16 verses yield the most detailed version of this story currently known to me, and it is therefore of great value.

It dates the incident as occurring 200 years ago (which, for reasons revealed later, suggests that the poem itself may date from around 1920), and features a man called McGloughlan who lived close to the shore of Glenade Lake, County Leitrim, with his wife, Grace Connolly.

GRACE CONNOLLY AND THE
DOBHAR-CHÚ OF GLENADE LAKE

One bright September morning, Grace visited the lake to bathe, but when she did not return home her husband retraced her steps, and upon reaching Glenade he found her dead body, torn and blood-stained—with her murderous assailant, the *dobhar-chú*, lying across her bosom. Maddened with grief and rage, he raced home for his gun, returned to the scene of the horrific crime, and shot his wife's killer dead. In the fleeting moments before it died, however, the *dobhar-chú* gave voice to a single piercing squeal—which was answered from the depths of the lake. Seconds later, the dead creature's avenging mate surfaced, and the terrified McGloughlan fled.

Reaching home, he told his neighbors what had happened, and they advised him to flee the area at once. This he did, accompanied by his loyal brother, both riding speedily on horseback, but doggedly pursued by the whistling *dobhar-chú*. After 20 miles, they reached Castlegarden Hill, dismounted, and placed their horses lengthwise across the path leading into it. Standing nearby, with daggers raised, they awaited the arrival of their shrill-voiced foe—and as it attempted to dash through the horses' limbs, McGloughlan plunged his dagger downwards, burying it up to its hilt within the creature's heart.

A COUPLE OF *DOBHAR-CHÚ* MONUMENTS

Needless to say, it would be easy to dismiss the story of Grace Connolly as nothing more than an interesting item of local folklore—were it not for the existence of two *dobhar-chú* tombstones commemorating the above episode. These are documented in an article by Patrick Tohall, published by the journal of the Royal Society of Antiquaries of Ireland in 1948. The first of the two monuments is a gravestone in Congbháil (Conwall) Cemetery in the town of Drumáin (Drummans), forming part of the approach to the Valley of Glenade from the coastal plain of north County Leitrim and south County Donegal, and just a few miles south of Kinlough, beside the main road leading from Bundoran to Manorhamilton.

A recumbent flag of sandstone roughly four feet, six inches by one foot, ten inches and dated September 24, 1722, what makes this the more interesting of the two stones is that it actually portrays the *dobhar-chú* itself, described by Tohall as follows:

THE DOBHAR-CHÚ (MASTER OTTER)
(Daev Walsh)

The carved figure is set in a panel about 17 1/2 ins.
by 7 ins. It shows a recumbent animal having body
and legs like those of a dog with the characteristic
depth of rib and strength of thigh. The tail, long
and curved, shows a definite tuft. The rear of the
haunch, and still more the tail, are in exceptionally
low relief, apparently due to the loss of a thin flake
from the face of the slab. So far the description is
canine. The paws, however, appear unusually
large, while the long, heavy neck and the short
head into which it shades off, together with the tiny
ears are all like those of an Otter or such Mustelida.

The head and neck are bent backward to lie flat
on the animal's backbone. A human right hand,
clenched and with fingers facing the spectator, is
shown holding a weapon which has entered the
base of the neck and reappears below the body in

a short stem which suddenly enlarges to finish as
a barb.

The article contains a photo of this depiction taken by society member Dr. J.J. Clarke. Unfortunately, in my files' photocopy of
Tohall's article, the illustrations had not reproduced well. In
autumn 1997, however, after I had communicated with *Fortean Times*
correspondent Daev Walsh concerning it, he and a colleague, Joe
Harte, independently visited the *dobhar-chú* gravestone in autumn
1997. Not only were they both able to confirm that it still existed,
they also took some excellent photographs of it, which they kindly passed on to me. These lucidly portray the carved *dobhar-chú*,
revealing that its head is indeed small and somewhat lutrine.
Equally, after studying the photos, I agree with Tohall's description
of its body as canine—almost greyhound-like, in fact, except for its
large paws and lengthy neck.

Interestingly, when I showed these pictures to various cryptozoological colleagues, some of them mistakenly assumed that the
clenched hand of the *dobhar-chú's* slayer was actually the creature's
head! However, it is far too small to be this, and when the photos
are viewed closely, the fingers of the clenched hand, which face the
camera, can be clearly discerned gripping a spear-like weapon, as
can the creature's real head, thrown backwards across its back. Even
the thin line of its mouth is readily visible.

Some of the wording on the gravestone is still legible too, identifying the person buried beneath as Grace Con, wife of Ter
MacLoghlin. According to Tohall, she was still spoken of locally, but
as Grainne, not Grace, and he also pointed out that Ter is undoubtedly short for Terence, and that it is Gaelic custom for a married
woman to retain her maiden name, which explains why Grace was
referred to on her gravestone as Con rather than MacLoghlin. Tohall
considered it likely that her gravestone was prepared while her
death was still fresh in local memory, because similar gravestones
in this same cemetery are characteristic of the period 1722 to 1760.

This, then, would appear to be the last resting place of the hapless young woman killed by the *dobhar-chú,* whose own existence is commemorated here too—all of which seemingly elevates the episode from folklore into fact.

As recently as World War I, the second *dobhar-chú* tombstone, believed to be that of Grace's husband, Ter, was still in the cemetery of Cill Rúisc (Kilroosk), at the southern entrance to Glenade, but had broken into two halves. At some later date, these were apparently placed up onto a boundary wall, and subsequently disappeared. Fortunately, however, at the time of Tohall's research it was still well remembered by all of the region's older men, who stated that it depicted some type of animal, and was popularly known as the Dobhar-Chú Stone. When asked whether the animal had resembled a dog, the only person who could recall the creature's appearance stated that it was more like a horse.

Recalling the story of the *dobhar-chú* in his article, Tohall placed the home of Grace (or Grainne) and her husband in the townland of Creevelea at the northwest corner of Glenade Lake, and (like L.A. Walkington, above) stated that Grace visited the lake to wash some clothes (not to bathe, as given in the 16-verse poem). Indeed, several variations of the story exist elsewhere in the general vicinity of Glenade, but Tohall believed that the Conwall tombstone was particularly important—for constituting possibly the only tangible evidence for the reality of the *dobhar-chú,* which, incidentally, appears in ancient Scottish lore too.

The Isle of Skye, for example, in Scotland's Inner Hebrides, is said to harbor a larger-than-normal type of otter dubbed "the king of the otters," with a white spot on its breast. It is only rarely spied and is very difficult to kill, but the pelage from one of these beasts is claimed to guarantee victory in battle to anyone who possesses it.

WHAT'S IN A NAME?

Tohall also offered some interesting reflections upon the terminology of the master otter's native name. Both in Ireland and in

Scotland, *"dobhar-chú,"* which translates as "water-hound," has two quite different meanings. One is merely an alternative name for the common otter *Lutra lutra*, but is rarely used in this capacity nowadays (superseded by *"mada-uisge"*). The other is the name of a mythical otter-like beast, and is still widely used in this capacity within the County Leitrim region. Tohall reserved the most intriguing insight into the master otter concept, however, for the closing sentence of his article:

> The best summary of the idea is set out in the records of the Coimisium le Béaloideas by Sean ó h-Eochaidh, of Teidhlinn, Co. Donegal, in a phrase which he heard in the Gaeltacht: "the Dobharchú is the seventh cub of the common otter" (*mada-uisge*): the Dobhar-chú was thus a super otter.

Another unusual creature, and name, apparently relating to the *dobhar-chú* is the so-called "Irish crocodile." A terrifyingly close encounter with one of these mystery beasts allegedly occurred at the six-mile-long Lough Mask (Measg), in County Mayo, western Ireland, in or around 1674, and was documented by Roderick O'Flaherty in his book *A Chorographical Description of West or H-Iar Connaught* (1684) as follows:

> Here is one rarity more, which we may terme the Irish crocodil [sic], whereof one as yet living, about ten years ago [1674], had sad experience. The man was passing the shore just by the waterside, and spyed [sic] far off the head of a beast swimming, which he tooke to have been an otter, and tooke no more notice of it; but the beast it seems there lifted up his head, to discern whereabouts the man was; then diving, swom [sic]

under water till he struck ground; whereupon he
runned out of the water suddenly, and tooke the
man by the elbow, whereby the man stooped
down, and the beast fastened his teeth in his pate,
and dragged him into the water; where the man
tooke hold on a stone by chance in his way, and
calling to minde he had a knife in his pocket,
tooke it out and gave a thrust of it to the beast,
which thereupon got away from him into the
lake. The water about him was all bloody,
whether from the beast's bloud [blood], or his
own, or from both, he knows not. It was of the
pitch of an ordinary greyhound, of a black slimy
skin, without hair, as he immagined [sic]. Old
men acquainted with the lake do tell there is such
a beast in it, and that a stout fellow with a wolf
dog [Irish wolfhound?] along with him met the
like there once; which after a long strugling [sic]
went away in spite of the man and dog, and was
a long time after found rotten in a rocky cave of
the lake, as the water decreased. The like, they
say, is seen in other lakes of Ireland; they call it
Dovarchu, i.e. a water-dog, or *Anchu*, which is the
same.

As can be seen from this description, Lough Mask's "Irish crocodile" certainly compares very closely with the Glenade *dobhar-
chú*—especially if one assumes that the former beast was not
actually hairless, but instead possessed short hair, which, when wet,
adhered so closely to its body that it seemed to its human victim to
be shiny and hairless (as is also true with normal otters). Peter
Costello documented this report in his own book, *In Search of Lake
Monsters* (1974), and he too believes that the *dobhar-chú* and Irish
crocodile are one and the same creature.

MODERN-DAY MEETINGS WITH
THE *DOBHAR-CHÚ?*

Today, the world beyond Glenade seems to have forgotten all about the *dobhar-chú* and its sinister deeds, but I wonder whether it may be premature for cryptozoology to assume that this mysterious animal is entirely confined to the shadows of the distant past—for it may conceivably have made some unexpected appearances as recently as 1968.

County Sligo (origin of a story featuring a horned specimen of *dobhar-chú*) is sandwiched between County Leitrim to the right, and County Mayo to the left. Several islands are present off the western coast of Mayo, including Achill, whose principal cryptozoological claim to fame is Sraheens Lough, a small lake said to be frequented by strange water monsters. Most of Ireland's examples are of the "horse-eel" variety—sinuous entities with horse-like heads, which may conceivably be undiscovered modern-day representatives of those supposedly long-extinct serpentine whales known as zeuglodonts.

The Sraheens Lough monster, however, is very different. On the evening of May 1, 1968, two local men, John Cooney and Michael McNulty, were driving past this lake on their way home when suddenly an extraordinary creature, shiny dark-brown in color and clearly illuminated by their vehicle's headlamps, raced across the road just in front of them and vanished into some dense foliage nearby. They estimated its height at around two and a half feet and its total length at eight to ten feet, which included a lengthy neck, and a long sturdy tail. It also had a head that they variously likened to a sheep's or a greyhound's, and four well-developed legs, upon which it rocked from side to side as it ran. Just a week later, a similar beast crawled out of the lake as 15-year-old Gay Dever was cycling by. It seemed to him to be much bigger than a horse, and black in color, with a sheep-like head, long neck, tail, and four legs (of which the hind pair were the larger). Other sightings were also reported during this period, but the identity of the animal(s) was never ascertained.

In the 13-volume encyclopedia *The Unexplained*, edited by Peter Brookesmith, and first published in part-work form during the early 1980s, an artist's reconstruction of the Sraheens Lough monster appeared in an article on Irish lake monsters written by Janet and Colin Bord (reproduced in a later compilation, *Creatures From Elsewhere*, in 1984). The illustration was based upon the eyewitness accounts given above for this monster, but it also happens to be exceedingly similar to the Conwall gravestone's depiction of the *dobhar-chú*. Both share a sleek body and powerful hindquarters, long slender tail, lengthy but not overly elongated neck, distinct paws, and relatively small head. Indeed, one could easily be forgiven for assuming that the two illustrations had been based upon the same animal specimen (let alone species). Could such an arresting degree of morphological similarity be nothing more than a coincidence, or is the *dobhar-chú* still in existence amid the countless lakes of the Emerald Isle?

Cryptozoological skeptics have pointed out that with a circumference of only 1,200 feet, surely Sraheens Lough is too small to support water monsters. However, anything capable of running across roads on four sturdy limbs is equally capable of moving from one lake to another, not residing permanently in any one body of water—which could explain why sightings of monsters in Sraheens Lough are sporadic rather than regular.

Judging from the data presented in this coverage, if the *dobhar-chú* is a real animal it is clearly much more distinctive than just an extra-large version of the normal otter — and the sea otter *Enhydra lutris* is an exclusively marine species that never reaches British coasts — but until a specimen is obtained (if ever), it will continue to linger with leprechaun-like evanescence amid the twilight limbo between Celtic folklore and contemporary fact.

CHAPTER 2

Tracking Mongolia's Death Worm—The Gobi Desert's Most Shocking Secret!

A...case worthy of note is that of Isadora Persano, the well-known journalist and duellist, who was found stark staring mad with a match-box in front of him which contained a remarkable worm, said to be unknown to science.

SIR ARTHUR CONAN DOYLE—"THE PROBLEM OF THOR BRIDGE," IN *THE CASE-BOOK OF SHERLOCK HOLMES*

WHILE REMAINING FORTUNATE ENOUGH NEVER TO have encountered an unidentified beast as deadly as the make-believe monster cited above, during my many years of cryptozoological research and documentation I have publicized a number of mystery creatures that were hitherto largely or wholly unknown even to the cryptozoological community.

To my mind, however, by far the most extraordinary member of this eclectic collection of cryptozoological newcomers is indeed a monstrous worm—a certain venomous vermiform with a truly electrifying presence...in every sense of the word!

Those nomadic tribesmen who share its sandy domain in the southern expanse of Mongolia's vast Gobi Desert refer to it as the *allghoi khorkhoi* (also variously transliterated into English as *allergorhai horhai* and *olgoj chorchoj*). Conversely, those few Westerners who have heard tell of it have very appropriately dubbed it the

Mongolian death worm.

According to the nomads' testimony, this astonishing animal can not only squirt a lethal corrosive venom, but also kill people merely by touching them (directly or indirectly), in a mysterious yet instantaneously fatal manner highly suggestive of electrocution!

This present chapter is an abridged version of my article "Meet Mongolia's Death Worm—The Shock of the New," published in *Fortean Studies* in 1998, which is the most detailed account ever published in relation to one of cryptozoology's latest, and greatest, enigmas.

PART 1:
DEATH WORM DATA AND DOCUMENTATION

MAKING MY ACQUAINTANCE WITH THE
ALLGHOI KHORKHOI

I first learned of the Mongolian death worm in 1995, when, while perusing a *World Explorer* issue from 1994, I came upon an article titled "In Search of the Killer Worm of Mongolia," authored by Ivan Mackerle. It described the first of two Gobi expeditions led by him during the 1990s in quest of the *allghoi khorkhoi*—a creature that seemed so utterly fantastic that at first I wondered whether the entire article was a hoax, especially as there were no accompanying biographical details regarding its author, or even any expedition photos.

In June 1996, however, a more detailed account by Ivan Mackerle was published in *Fate,* along with background information concerning him, as well as photographs depicting his search. The biographical data revealed that he was an author from Prague who has investigated a number of mysteries and regularly appeared on Czech television and radio. And when I pursued further lines of enquiry, I discovered that he was also a qualified design engineer and a well-respected explorer. In short, Ivan Mackerle was evidently a responsible researcher whose publications could be taken seriously.

Accordingly, I began communicating with Ivan by telephone and letter concerning the death worm and other cryptids. During our continuing correspondence, Ivan has very kindly made available to me for use in my own writings a veritable archive of material, most of which had never previously been cited in any English-language publication, and which comprises a considerable proportion of the data presented here. He also generously provided all the English translations of the various excerpts from Czech, Russian, and Mongolian sources quoted here or in my original *Fortean Studies* article.

DEFINING THE DEATH WORM

Somewhat unappetizingly, the death worm earns its native name from its alleged resemblance to an animate cow's intestine—"*allghoi khorkhoi*" translates as "intestine worm." A succinct account of this grotesque creature is provided in a detailed information sheet compiled by Ivan in July 1997, which assimilates data gathered during his expeditions and extensive bibliographical research. In the English translation of his information sheet, the death worm's "biography" reads as follows:

> Sausage-like worm over half a metre [18 inches] long, and thick as a man's arm, resembling the intestine of cattle. Its tail is short, as [if] it was cut off, but not tapered. It is difficult to tell its head from its tail because it has no visible eyes, nostrils, or mouth. Its colour is dark red, like blood or salami...It moves in odd ways—either it rolls around or it squirms sideways, sweeping its way about. It lives in desolate sand dunes and in the hot valleys of the Gobi desert with saxaul plants underground. It is possible to see it only during the hottest months of the year, June and July; later it burrows into the sand and sleeps. It gets out on [top of] the ground [i.e. sand] mainly after the rain, when the ground is wet. It is dangerous, because it can kill people and animals instantly at a range of several metres.

> Area of its appearance: Galbin Gobi, Dzagsuudzh Gobi, Ondoer Gobi, Jungaria Gobi, Khaldzan Uul, Orag Nuur, Khoerkh, Khoet. It is in the south part of Mongolia near the Chinese border.

ARTIST'S REPRESENTATION OF MONGOLIAN DEATH WORM
(Ivan Mackerle)

However, certain amplifications and discrepancies regarding its morphology should be noted. Some of the nomads interviewed by Ivan and the other members of his expeditions have asserted that the death worm can grow up to five feet long, and that it bears darker spots or blotches upon its red skin, which is smooth and unscaled. Moreover, one elderly woman who claimed to have seen a death worm when she was a girl stated that "it was sort of bound into points at both ends." This contradicts the earlier-noted description of the worm's posterior end as terminating abruptly rather than tapering, and may even indicate that the creature bears one or more pointed projections at both ends.

THE GOBI DESERT—
DOMAIN OF THE DEATH WORM

Home to the death worm is the Gobi, one of the great deserts of the world. Encompassing an area of approximately half a million square miles, it spans roughly a thousand miles from east to west across southeastern Mongolia and northern China, extending from the Great Khingan Mountains to the Tien Shan. It stretches about

500 miles from north to south.

Located on a plateau ranging 3,000 to 5,000 feet in height, the Gobi consists of a series of shallow alkaline basins, and its western portion is almost entirely sandy. Notwithstanding this, it supports a thriving diversity of wildlife, and its grassy margins are inhabited by nomadic Mongol tribes living as shepherds and goatherds. Many significant dinosaur remains, including fossilised eggs of the ceratopsian *Protoceratops*, have been discovered amid this region's vast expanse since the 1920s.

Whereas the word "desert" typically conjures up images of a hot, burning sandscape, this is in fact true only for certain deserts—those sited at low latitudes, such as the Sahara, caused by high-pressure air masses preventing precipitation. In contrast, the Gobi, of quite high-latitude location, is what is known as a cold desert, related to mountain barriers that deter moist maritime winds and thereby severely limit rainfall. The surface temperature in cold deserts can plummet so dramatically during the winter months that some of their fauna must remain beneath the surface of the sand in order to stay warm enough to survive, whereas others migrate to milder regions.

THE MACKERLE EXPEDITIONS

The first expedition took place during June and July 1990, the year's two likeliest months for encountering the worm on the surface, rather than beneath, the desert sands. The team was comprised of the following Czech members: Ivan Mackerle, the leader; Dr. Jarda (Jaroslav) Prokopec, physician; George (Jiří) Skupien, photographer; and Zdeněk Kropáč. Its two Mongolian members were Sugi, the translator; and Tschimed, the driver.

This was the first expedition ever launched specifically to search for the death worm. Even in Mongolia, it had remained largely unknown beyond its sandy seclusion—due in no small way to the country's former communist government prohibiting entry into many areas near to the Chinese border, coupled with an absence of public transport and private off-road vehicles. Mackerle's expedition chose to explore Galbin Gobi (around Khanbogd) and the areas extending to the west and southwest from Dalandzadgad, where it is reputed to exist. These regions included zones not previously visited even by Mongolian explorers.

The team had originally planned to entice some death worms out of the sand and into view with baited traps containing this species' favorite food, but none of the nomads could tell them what that was. Consequently, they had no option but to attempt an alternative approach—luring the worms with very intensive ground-borne vibrations. The initial plan was to generate these vibrations with a specially created "thumper" featuring a heavy weight and spring, and driven by a battery-run motor. Eventually, however, they abandoned this in favor of an ordinary wooden log, which was forced down into the ground on several occasions to create a succession of vibrations emanating outwards beneath the sand. This approach was inspired by Frank Herbert's classic science fiction novel, *Dune*, featuring a species of giant vibration-sensitive sandworm, the *shai-hulud*, inhabiting the desert planet Arrakis.

Unfortunately, neither the timing of their expedition during June and July nor their efforts to encourage their quarry's appearance upon the surface of the sand via vibrational stimuli achieved success; they did not encounter the death worm. Even so, they did succeed in collecting a very considerable amount of anecdotal evidence and testimony from local people, which had not previously been made available to European scholars.

Expedition #2 occurred in June and July 1992, and comprised Ivan, Jarda, and George, plus Tschimed as their Mongolian driver, and also a local guide called Nerguj. This time the team explored other portions of the area west from Dalandzadgad, which seemed from locals' accounts to offer a greater chance of success for encountering a death worm. They also interviewed lamas and shamans living close to the border with China, and set off several controlled detonations of explosives in the hope that the resulting shock waves traveling through the sands would induce any worms in the vicinity to emerge onto the surface. Sadly, however, the worms were not for turning—upwards, that is. Undeterred, the team filmed their search, which was broadcast as a 30-minute TV documentary entitled *Záhada Písečného Netvora* ("The Sand Monster Mystery") by Prague's Česká Televize in 1993.

A third Czech foray—"Olgoj Chorchoj Expedition 1996"— spanned 14 days in June 1996. This did not include Ivan, but featured two of his acquaintances—Miroslav Náplava and Petr Horký—plus two other Czech team members. Unlike Ivan's explorations, this one was not concerned solely with the death worm, so only a portion of the fortnight was allocated to seeking it. No firsthand sightings were made by the team, but they did obtain an interview with a local death worm eyewitness.

BEHAVIOR AND LIFESTYLE
OF THE DEATH WORM

During Ivan's field and bibliographical investigations, several significant and sometimes truly astonishing aspects concerning the death worm's behavior and lifestyle were uncovered. These can be itemized as follows:

> 1) There is a high frequency of death worm sightings in areas sustaining *Cynomorium songaricum*. This is a cigar-shaped, poisonous parasitic plant locally termed *goyo*, found on the saxaul plant's roots, which are also poisonous.

> 2) One old woman, called Puret, claimed that when the worm attacks, it raises half of its body up through the sand, inflates itself, and secretes a bubble of poison from one end, ultimately squirting it forth in a stream at its unfortunate victim.

3) Puret also claimed that anything contacted by this deadly fluid turns yellow instantly, and looks as if it had been corroded by acid. However, it loses its potency from the end of June onwards, and meeting the worm then does not always result in death.

4) Ivan and his team repeatedly heard claims from the nomads that the death worm can kill people by touch, too (and sometimes even when several feet away). While camping near a monastery, the team was informed by their on-site guide, Khamgalagu, hailing from Khanbogd, about a local boy who had accidentally touched a death worm that had found its way into a box in his parents' campsite and had concealed itself within it. The boy died instantaneously — as did his parents when they attempted to kill it.

5) During Ivan's first expedition, his team's interpreter, Sugi, recalled an incident from his childhood when a party of geologists were visiting his home region. One of the geologists had been idly poking the sand with an iron rod when suddenly, without any warning, he dropped down onto the ground. His horrified colleagues raced over to him at once, but he was dead. As they peered at the sand that he had been prodding, however, it began to churn violently, and out of it emerged a huge fat worm—an *allghoi khorkhoi*. Yet the man had not touched this deadly creature directly, only via the metal rod.

6) The team also spoke with Yanzhingin

Mahgalzhav, a nature ranger from
Dalandzadgad, who affirmed (as did Puret) that
during the 1960s a single death worm had killed
an entire herd of camels just south of Noyon,
when they unsuspectingly plodded across an
expanse of sand concealing one of these dreaded
beasts lying beneath the surface.

Not surprisingly, the Gobi nomads greatly fear this apparently lethal creature and take great care not to draw close to it if they happen to encounter one unawares.

The hypotheses generated by Ivan and his colleagues from the above data are exceedingly thought provoking. For example: they consider it possible that the reason for the death worm's close association with the *goyo* plants is that it obtains its venom from them, or from the roots of their host, the saxaul. Even more extraordinary, if true, is the prospect that the death worm can kill by electrocution. Although a highly radical proposal, which I shall discuss fully in Part 2 of this chapter, it could certainly explain the instantaneous deaths attributed to this worm of people who have touched it or camels that have unsuspectingly trodden upon it. Moreover, assuming Sugi's anecdote is true, electrocution undeniably offers an effective mechanism by which someone could die instantly even after having touched the worm not directly but with an iron rod — an excellent electrical conductor.

DEATH WORM DOCUMENTATION

Although Ivan Mackerle is unquestionably the most famous of the Mongolian death worm's investigators and chroniclers today, he is by no means the first. A comprehensive chronology is presented in my *Fortean Times* article, whose most significant entries are as follows.

The earliest specific reference to the death worm that I have so far obtained is an excerpt from *On the Trail of Ancient Man* (1926), written by the eminent American paleontologist Prof. Roy Chapman

Andrews. This book concerns the American Museum of Natural History's famous Central Asiatic Expedition of 1922 to the Gobi, led by Prof. Andrews, in search of dinosaur fossils. In order to obtain the necessary permits to venture forth into the Gobi, Andrews needed to meet the Mongolian Cabinet at the Foreign Office.

When he arrived, he discovered that numerous officials were in attendance for their meeting, including the Minister of Foreign Affairs and the Mongolian Premier himself. After Andrews had signed the required agreement in order to obtain the expedition's permits, the Premier made one final but very unusual and totally unexpected request:

> Then the Premier asked that, if it were possible, I should capture for the Mongolian government a specimen of the *allergorhai-horhai*. I doubt whether any of my scientific readers can identify this animal. I could, because I had heard of it often. None of those present ever had seen the creature, but they all firmly believed in its existence and described it minutely. It is shaped like a sausage about two feet long, has no head nor legs and is so poisonous that merely to touch it means instant death. It lives in the most desolate parts of the Gobi Desert, whither we were going. To the Mongols it seems to be what the dragon is to the Chinese. The Premier said that, although he had never seen it himself, he knew a man who had and had lived to tell the tale. Then a Cabinet Minister stated that "the cousin of his late wife's sister" had also seen it. I promised to produce the *allergorhai-horhai* if we chanced to cross its path, and explained how it could be seized by means of long steel collecting forceps; moreover, I could wear dark glasses, so that the disastrous effects of

even looking at so poisonous a creature would be
neutralized. The meeting adjourned with the best
of feeling.

Call me a cynic, but I have the distinct impression that Prof.
Andrews did not take the death worm too seriously. In any event,
he certainly didn't succeed in finding one, which is probably no bad
thing—bearing in mind that he had planned to pick up with steel
forceps a creature that had allegedly killed a fellow geologist who
had prodded it with a metal rod!

During the 1920s, the American Museum of Natural History
sent forth several additional Central Asiatic Expeditions to
Mongolia and China, and in 1932 a major work, *The New Conquest
of Central Asia,* was published, documenting all of them, with Prof.
Andrews as its principal author. The first volume in the series
Natural History of Central Asia (edited by Dr. Chester A. Reeds) contained a brief section titled "The Allergorhai Horhai":

> At the Cabinet meeting the Premier asked that I
> should capture for the Mongolian Government a
> specimen of the *Allergorhai horhai.* This is proba-
> bly an entirely mythical animal, but it may have
> some little basis in fact, for every northern
> Mongol firmly believes in it and will give essen-
> tially the same description. It is said to be about
> two feet long, the body shaped like a sausage, and
> to have no head or legs; it is so poisonous that
> even to touch it means instant death. It is reported
> to live in the most arid, sandy regions of the west-
> ern Gobi. What reptile can have furnished the
> basis for the description is a mystery!
>
> I have never yet found a Mongol who was will-
> ing to admit that he had actually seen it himself,

although dozens say they know men who have. Moreover, whenever we went to a region which was said to be a favorite habitat of the beast, the Mongols at that particular spot said that it could be found in abundance a few miles away. Were not the belief in its existence so firm and general, I would dismiss it as a myth. I report it here with the hope that future explorers of the Gobi may have better success than we had in running to earth the *Allergorhai horhai*.

Regrettably, however, as Ivan Mackerle and his fellow investigators can readily testify, such success continues to elude those who seek this evanescent monster.

From 1946 to 1949, the U.S.S.R.'s Academy of Sciences launched a series of expeditions into the Gobi, led by Dr. Yuri Orlov. In 1946, Ivan Antonovich Efremov (=Yephremov), a Russian paleontologist, took part in one of these, which he later documented in his book *Doroga Vetrov* ("The Wind's Path") (1958). During the expedition, Danzang, a young geologist who could speak the local Mongolian dialect, was instructed to inquire about the death worm with an elderly man, Tseveng, from Dalandzadgad. Danzang himself had never heard of the death worm, so he expected to be harshly rebuffed or derided by Tseveng for wasting his time. Instead, to Danzang's great surprise, Tseveng informed him that he was indeed aware of this creature, and that although he had not spied it personally, he had been informed about it on many different occasions from other Gobi nomads.

According to Tseveng, the death worm could kill with a single stroke and inhabited a barren expanse of wasteland known as Khaldzan-dzakh, roughly 80 miles southeast of Dalandzadgad. Here it remained hidden beneath the sand dunes for most of the year, but surfaced during the summer months of June and July. When Danzang translated Tseveng's words, the Russian geologists began to joke about the death worm's alleged killing abilities, which greatly angered the elderly Mongol. Scowling with rage, he fired off a terse volley of words before abruptly departing—words which, when interpreted by Danzang, left the team in no doubt as to the very real fear engendered among the local people by this formidable beast:

> You laugh only because you know nothing and
> understand nothing. The *allghoi khorkhoi*—it is a
> terrible thing!

Efremov speculated that the Mongol nomads' death worm traditions and lore may derive from some "living fossil," perpetuating its line from prehistory into historic times—but which, if not already recently extinct, is now very rare, persisting only in the most remote areas of Central Asia.

Incidentally, preceding *Doroga Vetrov* was the publication in 1954 of *Stories* (translated into English by O. Gorchakov, and also appearing in French as *Récits*), a collection of science fiction tales. Indeed, Efremov is actually more famous as a SF writer than as a paleontologist. In the introduction to this collection, Efremov wrote:

> To try to lift the curtain of mystery over these
> roads, to speak of scientific achievements yet to
> come as realities, and in this way to lead the read-
> er to the most advanced outposts of science—
> such are the tasks of science-fiction, as I see them.

One of those "scientific achievements yet to come as realities" is of particular interest to cryptozoology, because among his stories in this collection is one titled "Olgoï-Khorkhoï." And sure enough, its theme is none other than the fatal discovery by a geological party of the Gobi's infamous death worm! Clearly inspired by local testimony that he had gathered during his own Gobi expedition, Efremov's tale contained a detailed account of the *allghoi khorkhoi,* corresponding with the reports documented elsewhere by Ivan Mackerle and others—including its lethal power of apparent electrocution.

In the story, two worms killed some of the party's members even without physically touching them, again echoing true-life nomad lore. However, an intriguing additional detail contained in Efremov's science-fiction story is that immediately before the worms killed their victims, they writhed in a spiral and changed color—becoming purple with bright blue tips. Could this be a facet of death worm behavior collected by Efremov during his Gobi expedition of 1946 but not noted in *Doroga Vetrov?*

In 1966, Y. Tsevel's notable dictionary of the Mongol language, *Mongol Khelnii Tovttch Tailbar Toli,* included an entry for the *allghoi khorkhoi.* This described it as a Gobi-inhabiting worm that resembles a huge intestine and is extremely venomous.

Speaking of dictionaries: Ivan Mackerle has revealed to me a fascinating snippet of information indicating that the death worm is known not only in Mongolia but also in Kazakhstan. Reading through a Mongolian-Kazakh dictionary, he discovered that the Kazakhs have their own term for the *allghoi khorkhoi*—namely, "büjenzhylan."

Altajn Tsaadakh Govd (1987) by Mongolian author Dondogijn Tsevegmid (=Cevegmid) documents the wide deserts and beautiful mountains lying to the south and southwest of the Altai Mountains—a vast region colloquially termed the Behind-Altai Gobi. It also provides several Altai-based cryptozoological snippets, including items mentioning Mongolia's famous man-beast, the *almas,* as well as a lesser-known version, the *chun goeroes.*

For our purposes, however, the most noteworthy section is one that deals with the death worm, and also with what may be a related yet equally unidentifiable mystery beast—not to mention a self-inflating hedgehog! It reads as follows:

At that time, a local herdsman from Noyon

sumyn [village] in Oemnoegov aimak [country] talked about the *shar khorkhoi* (yellow worm), *zamba zaraa* (hedgehog) and *allghoi khorkhoi* (intestine worm).

"I was riding my camel along the Tost mountains. Suddenly I spotted a long yellow animal in front of me. It looked like neither fox nor wolf...My camel started to cry out in fear, and its cry attracted others of the same creature, which began to come out of their holes in the ground and approach me. So I took flight with fear. When I turned my head while riding away, I saw about 50 of the creatures following me. Luckily for me, I met some nomads, riding against me. They told me that they were *shar khorkhois*, living underground. Lately they are rare.

"In the west part of the Segs Cagan Bogd mountains lives a strange animal, the *zamba zaraa*. I met one once while hunting. It sat on a big stone, with its head erect, which inflated. It smacked its tail on the stone with such force that the noise it made was as loud as the sound of a camel's gallop. Because I had heard that the *zamba zaraa* can reach the size of a yurta [big tent] by inflating, I decided to take flight."

In addition to these two strange animals, another, more dangerous animal also lives in the Gobi, the *allghoi khorkhoi*. It resembles an intestine filled with blood, and it travels underground. Its movement can be detected from above via the waves of sand that it displaces.

It is difficult to decide whether the *zamba zaraa* is beast or balloon! And it seems likely that the herdsman's account of the *shar khorkhoi* was enhanced by more than a little imaginative license. Nonetheless, we should not ignore the fact that whatever this creature is, it has its own specific local name, and I am not aware of any species whose morphology and activity recalls that of these peculiar yellow worms. As for the death worm, the description cited above corresponds well with the version given by many independent sources elsewhere.

Another Mongolian author who has alluded to the death worm is S. Dzhambaldorzh in *Mongol Nuucyn Chelchee* ("Braid of Mongolian Secrets") (1990), which includes a section titled "The most interesting rare worm in the world." This is of particular value, as it presents some information publicized in 1930 by Soviet scientist A.D. Simukov.

Simukov died while being transported to the Gulag, and all of his archives were supposedly destroyed. Clearly, however, some of his data concerning the death worm must have survived, which Dzhambaldorzh later uncovered and preserved for future generations by incorporating it within his own book:

> The *allghoi khorkhoi* is an animal that lives in Dzagsuudzh Gobi. The people there speak a great deal about it, and are afraid of it. It comes out onto the surface of the ground mainly after the rain, when the ground is wet. They say that its colour is white. The *allghoi khorkhoi* is a reptile unknown to the scientists, an amphibian, or a giant worm. It lives in hot hollows with saxaul plants, underground. The biggest is as thick as an arm, and about 1 m [three feet] long. The tail is short, as if it were cut off [i.e. truncated], but not tapered. The head is not distinguishable. This is what Soviet scientist A.D. Simukov wrote in 1930.

Much of this has since been substantiated by Ivan Mackerle's conversations with the nomads during his two expeditions, but there is one striking discrepancy—the worm's color. All other sources have claimed that the worm is red, but according to Simukov it is white. Is it possible that Simukov had inadvertently confused descriptions of some other species of animal with those of the death worm? There is, after all, the equally mysterious yellow worm or *shar khorkhoi* already on file.

Dzhambaldorzh added a further twist to this tale by including a very curious report concerning what appears to be a death worm with wings! Penned by geographer B. Avirmed from the Mongolian Geographical Institute, this report had first appeared in 1981 in a Mongolian newspaper. It featured the eyewitness testimony of a Mongolian shepherd:

> Shepherd L. Chorloo (Khorlaw) from Chongor
> Gobi in the south Gobi aimak (country) stated:
> "Here we see an interesting creature. Its body
> looks like salami, half of which is taken up by
> the head, and on the rear it has wings. I have
> seen it twice. On both occasions it was lying
> dead at the well."

In 1991, Avirmed and P. Tsolmon co-authored a more extensive article, in the Mongolian magazine *Shinzhlekh Ukhaan Amdral*. This included a sighting from 1982 of an unidentified beast in the Altai region of western Mongolia by B. Boldoos, a driver working for an expedition sent to the Chovd area by the Mongolian Academy of Science's Chemical Institute. The expedition's original driver, a man called Manalzhav, had become ill, and was replaced by Boldoos, who arrived by airplane. He was then collected by some of the expedition's members in a car and driven to the camp:

During the journey from the airport to the camp, they stopped at a place called Eezh Chairchan to have a rest and stay overnight. As they got out of the car and looked down at the ground, they saw a strange trace in the sand, forming right before their eyes. It looked as if something was moving under the sand, displacing a wave of sand above it. Out of the car stepped Boldoos's friend, Altanchuiag, who took a shovel, and when he scraped away the moving sand, a strange animal with spade-like paws jumped out and immediately buried itself again. It was beige in colour, was similar in length to [? - Ivan Mackerle was unable to translate this Mongolian word], and at the rear of its body it had paws or legs, but they did not move. The frightened men quickly withdrew.

Whatever Boldoos's sand-digging animal is, it bears little or no resemblance to the death worm, and exists not in the Gobi of southern Mongolia but in the Altai of western Mongolia. Consequently, I would not normally have included it in the present article—were it not for the perplexing testimony of the Mongol shepherd, whose "winged salami" beast constitutes a morphologically (albeit mystifyingly) intermediate creature between the "true" death worm and Boldoos's vaguely defined mystery beast.

The earliest of Ivan Mackerle's death worm articles in my files was published in 1991 by the Czech magazine *Reflex*, and another Czech magazine, *Filip*, published a similar article by Ivan in 1992. The earliest of Ivan's English-language death worm articles also appeared in 1992, within the September/October issue of *The Faithist Journal*, and was identical to his aforementioned *World Explorer* article from 1994.

The year 1993 saw the publication of the first cryptozoology-oriented book to include mention of the death worm. Entitled *Legendární Příšery a Skutečná Zvířata*, its author was Jaroslav Mareš, the Czech Republic's premier cryptozoologist, who devoted an entire chapter to the Gobi's cryptic denizen.

My own series of death worm writings began in 1995, with a section devoted to this cryptid appearing in the second of my "Menagerie of Mystery" columns for *Strange Magazine* (fall 1995). My second death worm article appeared in the March 1996 issue of *Wild About Animals,* summarizing the information that I had presented in more detailed form within my *Strange Magazine* account. It also included a superb color reconstruction by artist Philippa Foster (née Coxall) of the worm's likely appearance in two different poses—lying concealed beneath the sand but ready to discharge electricity, and in the act of squirting forth its deadly venom.

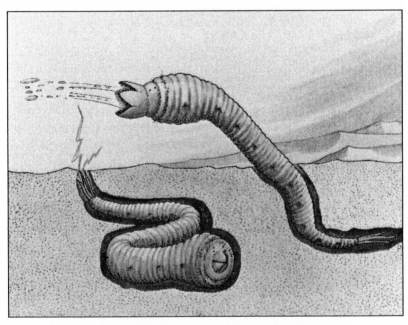

RECONSTRUCTION OF DEATH WORM BY PHILIPPA FOSTER
(Fortean Picture Library/Philippa Foster)

Ivan Mackerle's extensive *Fate* article was published in June 1996, which was essentially an expanded version of his *Faithist Journal/World Explorer* article, and contained a number of new illustrations, including two artistic reconstructions of the death worm, and a photo of the *goyo* plants. These appeared in black and white.

Luboš Koláček published a lengthy article in Prague's *Telegraf* newspaper on September 21, 1996, covering the "Olgoj Chorchoj Expedition 1996" featuring Miroslav Náplava and Petr Horký, and including some photos of the team members. As pointed out in the article, however, the death worm was not the only subject of interest to this particular expedition, which focused extensively upon the Gobi nomads themselves, as well as an erstwhile Mongolian cryptid—Przewalski's wild horse, undescribed by science until 1881.

Also published in September 1996, *The Unexplained: An Illustrated Guide to the World's Natural and Paranormal Mysteries* was the first of my books to refer to the death worm. Chapter 4, titled "Asia: The Occult and the Orient," contained a short account of Mongolia's cryptid, accompanied by Philippa Foster's full-color death worm picture.

A few weeks later, the death worm made its *Fortean Times* debut, via my article in *Fortean Times Weird Year 1996,* presenting a concise history of this beast. I titled it "Do Nomads Dream of Electric Worms?"—Philip K. Dick aficionados will explain!

An article by Jaroslav Mareš assessing the death worm appeared in 1996 within the Czech magazine *Mladý Svét,* as #25 in a series of articles dealing with cryptids. The title of the series, "Na Stopě Tajemných Zvířat," translates into English as "On the Track of Unknown Animals." Now where have I come across that title before?

In his account, Mareš recalled a very shocking anecdote—literally!—collected by him from a local herdsman while seeking dinosaur bones at Nemeght in 1967:

"My brother living in Obotó Chajun aimak knew a man who encountered an *allghoi khorkhoi*," one herdsman told me. "His name was Altan. Once he returned with a friend from a neighbouring camp. They were riding their horses, and it was just after noon, on a day in July. The sun was shining....

"Suddenly Altan's friend's horse fell down. The rider stood up and went to the horse, but suddenly cried out and fell again. Altan was 5 metres behind and saw a big fat worm slowly crawling away. Altan stood in horror and then ran to his friend. But he was dead, and so too was his horse."

Also on my files for 1996 is a detailed article by Ivan Mackerle concerning his expeditions that appeared on December 13, 1996, in the Czech publication *Společenství*. It included a computer-reconstructed image of the death worm's likely appearance, portraying it with a thick, extensively annulated (ringed) body and indistinguishable, featureless poles.

In February 1997, an in-depth article of mine documenting the death worm appeared in *Uri Geller's Encounters*, and included several of Ivan's color photos—depicting scenes from his expeditions, the *goyo* plants, and full-color reconstructions of the worm itself. I also presented an extensive analysis of the worm's feasible taxonomic identity.

In March, the death worm entered the rarefied realm of Britain's "quality newspapers." Jonathan Leake, the *Sunday Times'* environment correspondent, consulted with me in relation to a major feature that he was preparing on cryptozoology, and one of the cryptids whose details I supplied to him was the *allghoi khorkhoi*. His article was published on March 23, 1997, and contained a glorious full-color reconstruction of a triffid-lookalike death worm in the process of spewing forth its vitriolic venom. The report attracted notable media interest. The following morning, I spoke about the death worm live on Radio 4's *Today* program, and have publicized it on a number of other shows since then.

But that was not all. I was soon to learn that the devastating death worm had even penetrated cyberspace, with several websites devoted to it and countless references to it on others. Also in 1997, the death worm was the subject of a French Internet article by Michel Raynal. It was originally posted in his Institut Virtuel de Cryptozoologie website, but it was also published in a two-part hard copy form by the Belgian magazine *Cryptozoologia* in its June and July-August issues. (Throughout his article, Raynal referred to the death worm as the "intestine worm"—a direct translation of *"all-ghoi khorkhoi"*—but I consider this to be an unsatisfactory name, best avoided, because it readily invites confusion with tapeworms and other parasitic worms inhabiting the intestinal tract of their hosts.)

More recently, as the *Guinness Book of Records'* cryptozoological consultant, I was asked to suggest some cryptids that could readily slot into their superlative-structured format. One of those that I put forward for consideration was the death worm, and in the 1998 edition it duly appeared as "Most Dangerous Mystery Animal?"

In the years since then, numerous accounts concerning the death worm in hard-copy and online form have appeared, and this relative newcomer to the cryptozoological stage seems destined to inspire many more in the future.

PART 2:
UNMASKING THE DEATH WORM

With no physical evidence available for study, it is impossible to state categorically that the death worm is real. It may be entirely the product of native folklore and superstition. However, for the purpose of conducting a cryptozoological analysis of its taxonomic identity, we must assume that it is real—but I must reiterate that this is merely an assumption, not a fact. Yet even if we agree to suspend disbelief in this way, how similar is the "real" death worm to the version described by the nomads? Put another way, the death worm could be:

> (a) A real, but harmless species, with its various death-dealing capabilities merely the product of native folklore and fear;
> > or:

> (b) A real species equipped with the capability of emitting a highly toxic venom and/or able to generate and discharge electricity.

As for its taxonomic status, several options come to mind, and each will now be assessed.

WORMING AN IDENTITY OUT OF THE OLIGOCHAETES?

A profound problem with the word "worm" is that even in zoological parlance, let alone the general English language, it has been applied to a vast range of wholly unrelated animals—almost anything, in fact, that is long, relatively slender, limbless, and alive. Snakes, legless lizards, certain serpentine fishes, limbless insect larvae, and most of the invertebrate phyla have all at one time or another been referred to as worms. From a taxonomic standpoint, therefore, the word is wholly valueless as a source of clues regarding this cryptid's status.

Nevertheless, by comparing and contrasting the death worm's profile with that of each taxonomic group whose members loosely conform to the "worm stereotype" outlined above, we may indeed gain some pointers as to its most plausible identity.

Of the many invertebrate taxa referred to colloquially as "worms," only the segmented worms or annelids offer any realistic candidature in relation to the death worm's identity. Characterized by species whose bodies are composed of numerous ring-like segments and hence are also known as ringed worms (derived from "Annelida"), this phylum comprises three principal taxonomic groups—the polychaetes (including the familiar lugworms and ragworms), leeches, and oligochaetes. Of these, the only ones exhibiting any degree of similarity to the death worm are the oligochaetes—which include among their membership not only many freshwater species, such as *Tubifex*, but also the earthworms.

Could it be that the death worm is nothing more than a novel species of earthworm? Its elongated body and featureless poles—lacking an externally differentiated head or tail, and any visible sensory organs or mouth—certainly recall the basic earthworm configuration. So too does its fossorial lifestyle, and tendency to emerge onto the surface following rainfall. Even its strange mode of locomotion has been likened to the motion of a worm by Ivan Mackerle: "like a worm, it contracts and expands, or squirms to move about." This description encourages comparison with earthworms' familiar peristaltic movements, referred to in zoological parlance as vermiform locomotion.

Moreover, if we suppose that reports describing the death worm's ability to squirt poison at anyone approaching too closely are genuine, as this creature appears to lack eyes it must sense the arrival of persons, camels, and other creatures via their footsteps' vibrations—and sensitivity to terrestrial vibrations is a well-known characteristic of earthworms.

The death worm's fairly large size, ranging from 18 inches to five feet, according to local testimony, does not preclude an earthworm identity, either. One of Australia's most celebrated non-marsupial animals is its giant earthworm *Megascolides australis*, from Victoria, which can attain a total length of 13 feet. Nor is this the largest species on record. That honor goes to South Africa's *Microchaetus rappi*—one specimen of which, collected in 1936 by Van Heerden, measured a stupendous 22 feet!

Some species of earthworm also exhibit a remarkable talent more than a little reminiscent of one of the death worm's alleged abilities. Known aptly as squirter earthworms, they are typified by *Didymogaster sylvaticus*, which is native to New South Wales and able to defend itself by squirting jets of internal fluid for distances of up to 18 inches out of small pores surrounding its body. Although the fluid is totally harmless, its emission in this manner is more than adequate to startle and deter would-be predators.

Recalling the nomads' reports, could it be that the death worm is a dramatically new species of squirter earthworm, one that has developed a highly toxic squirting fluid? Although such a notion is undeniably radical, it is not impossible. Indeed, the death worm need not even be responsible for creating the toxin.

As noted earlier, Ivan Mackerle has suggested that the *allghoi khorkhoi* may obtain its venom from the poisonous roots of the saxaul, or from its parasite, the *goyo* plant, with which this mysterious animal is reputedly closely associated. If so, this is very interesting but hardly unprecedented. Many poisonous species of animal derive their toxins from external sources, including South America's famous arrow-poison frogs, whose skin contains deadly lipophilic alkaloid poisons (used by Indian tribes to tip their arrows), but which are now known to be derived from the ants that these amphibians eat.

Regardless of origin, however, the mystery remains as to how such a toxin would function, bearing in mind that it is supposed to be instantly fatal, yet is merely squirted by the worm at its victim—

rather than physically injected into it, as with venomous snakes or comparable predators. According to the old woman Puret, everything touched by this substance, even metal, looks as if it had been corroded by an acid. Always assuming, of course, that such a substance is indeed more than just a myth, perhaps, therefore, the death worm's expelled fluid is an exceedingly acidic, fast-acting toxin that achieves its deadly effect by searing its way through the victim's flesh and thence into its blood system.

Another apparent enigma associated with reports of the death worm's poison is the claim that it loses its strength from the end of June onwards, after which it is not always fatal. In fact, as the worm is usually encountered only during June and July anyway, this attenuation of the poison's potency is not so strange. After all, the worm never consumes any of its victims, so its poison has clearly not evolved as a means for obtaining food. Instead, it evidently functions as a defense mechanism, serving solely to protect the worm from would-be attackers or predators. Consequently, there is no evolutionary advantage to the worm in manufacturing a poison in lethal concentrations during the ten months when this species is usually hidden from danger beneath the desert sands.

Conversely, as a wholly alternative scenario, perhaps the death worm is a squirter earthworm whose fluid is indeed harmless after all—but because of its body's large robust form, its squirting effect is sufficiently dramatic to have nurtured a superstitious, erroneous fear among the Mongol nomads that it is actually extremely venomous. Even so, whereas squirter earthworms emit fluid all along their body, the death worm supposedly expels its deadly emission only from the tip of one end of its body—a notable discrepancy.

The nomads do not seem to have any knowledge regarding the death worm's diet, but, ironically, this ignorance might actually substantiate an earthworm identity for it. Like earthworms, the death worm may possess only a tiny, visibly insignificant mouth, and simply feed upon decaying organic matter. It may also ingest sand while burrowing (just as earthworms ingest soil), absorbing into its gut organic material present in the sand, then defecating the sand. It might also absorb nutrients directly through its skin.

Similarly, as it does not appear to have any visible nasal aperture(s), the death worm may respire through its skin too. Most oligochaetes do this, containing within their outer skin layer an extensive capillary network that often confers a bright red coloration upon their skin. Could this also explain the deep red coloration reported for the death worm?

These, then, are among the more positive features allying the death worm with the earthworms. However, there are also some shortcomings to take into account. The most serious of these stems from the death worm's habitat—the Gobi desert.

Such an environment is the absolute antithesis of the preferred domain of earthworms, which require moist, damp environs due to their permeable body surface. Earthworms placed in arid environments would rapidly dry out and die, unless a species could evolve that displayed the fundamental requirement for invertebrate survival here—an impervious external cuticle. Spiders, scorpions, and other desert-dwelling arthropods all possess a cuticle of this nature, but as yet no such adaptation has been recorded from oligochaetes.

However, there is one mitigating factor. Certain earthworms do live in relatively dry soils, and are able to avoid excessive water loss through their skin via the development of specialized excretory organs known as enteronephric nephridia. Instead of excreting urine directly into the outside world, these modified nephridia pass the urine into the digestive system, where most of the water can be reabsorbed while going through the intestine. Such a system, especially if enhanced by further evolution, would be invaluable to any earthworm attempting to survive in the desert.

Even so, another contradiction between the death worm's behavior and that of bona fide earthworms is that it only surfaces during the Gobi's two hottest months, June and July. Earthworms, conversely, tend to burrow deeper into the soil during hot, dry periods, sometimes entering a period of quiescence or aestivation until damper conditions return. Unless the death worm is an earthworm that has developed an impervious cuticle, it is difficult to reconcile its sunbathing proclivity with anticipated earthworm activity.

AN EXPLOSIVE ENIGMA FROM KALMYKIA

I mentioned earlier that Ivan Mackerle discovered that the people of Kazakhstan have their own name for the *allghoi khorkhoi*. In addition, a very curious type of vermiform mystery beast that may (or may not) be allied to the death worm has also been reported from the steppes and desert dunes of Kalmykia. This is a region of Russia to the north of Chechnya and Dagestan, and lies immediately to the west of Kazakhstan.

According to a letter of January 6, 1997, written to Michel Raynal by veteran Russian cryptozoologist Dr. Marie-Jeanne Koffmann, this unidentified creature is referred to by the Kalmyks as the "short grey snake." Measuring 50 cm (20 inches) long and 15 to 20 cm (six to eight inches) in diameter, it has smooth grey skin, and is rounded at its anterior end, but terminates abruptly with a very short tail. So far, its local "snake" appellation would seem to be appropriate, but it also has one characteristic that instantly sets it apart from any bona fide serpent and ostensibly places it among vermiforms of the invertebrate kind instead — for according to the Kalmyks, their so-called "short grey snake" does not possess any bones.

This would appear to be substantiated by their claim that if one of these beasts is struck hard in the middle of its back with a stick, it explodes—leaving behind a patch of slime or grease stretching more than a meter in diameter across the ground as the only evidence of its former existence. Although she is not absolutely certain (her original notes were destroyed during a burglary in her office), Dr. Koffmann believes she was told that this animal is slow moving, and moves in a worm-like manner. As to whether it is dangerous, however, some Kalmyks affirm that it is, but others state that it is not.

No mention is given of any facial features (although Koffmann claims that a second, smaller variety also exists here, which has a clearly delineated mouth). In any event, Kalmykia's exploding "worm" exhibits sufficient differences from the *allghoi khorkhoi* for me to see little reason for assuming that these two creatures share anything other than the dubious honor of being presently unrecognized and thus ignored by modern-day science.

SEALED SKINS AND CAECILIANS

Also called gymnophionids, caecilians are limbless soil-burrowing amphibians whose slender, externally annulated bodies with confusingly similar head and tail (when present) afford them a striking outward resemblance to earthworms—so much so that they are easily mistaken for them when seen. Having said that, caecilians are rarely seen, because they generally emerge onto the soil surface only at night, or if washed out of their underground burrows by rain (there are also a few genuinely aquatic species, and some others have aquatic larvae). When moving, they often undergo vermiform locomotion, just like bona fide earthworms.

The caecilian head is very nondescript, equipped with an inconspicuous mouth with which it seizes earthworms, arthropods, and small soil-dwelling vertebrates as prey, plus a pair of laterally sited eyes. These are tiny and usually functionless (sometimes concealed beneath the skin, or even underneath skull bones in some species). There is also a protrusible tentacle housed in a small pit on each side of the head. Although a caecilian's body looks smooth, it often possesses small calcified scales embedded within its skin.

The world's largest known species is Thompson's caecilian *Caecilia thompsoni* from Colombia, which measures approximately four and a half feet long, but most of the other 70-odd species presently recognized are under two feet in length. None has a diameter exceeding two inches.

Caecilians are confined to warm temperate and tropical regions in Central Africa, Central and South America, India, and southeastern Asia. On account of their subterranean lifestyle, however, they are easily missed, and new species are still being discovered. Could the death worm thus be an unknown species of sizeable caecilian? Its morphology is certainly similar, but its arid, sand-swept desert terrain is very different from the moist, damp, tropical soil world inhabited by those caecilians currently documented.

If a caecilian were to survive in what, for these amphibians, would be as atypical and extreme an environment as a desert, it would need to develop a very efficient means of avoiding desiccation. Such an adaptation would not be unprecedented among amphibians, however, for there are a number of desert-dwelling frogs and toads. These avoid the perils of water loss by a variety of different means.

Some, for instance, bury themselves deep in the sand and undergo periods of dormancy or aestivation, ensheathed in a cocoon of hardened mucus, or composed of several layers of dead skin formed from unshed epithelial cells. Certain other desert amphibians possess a layer of skin rich in mucopolysaccharides, which "seals in" their body water by acting as a sponge, binding water and releasing it only when required. Even so, the image of a caecilian lying on the surface of the sand during June and July, the Gobi's two hottest months, as the death worm is said to do, remains for me a very difficult one to accept.

Also, the death worm's toxin-squirting capability cannot be explained by the caecilian identity at all, because these apodous amphibians are neither toxic nor able to emit fluid in this manner. However, as with many other harmless animals, some native people firmly if erroneously choose to believe otherwise. For instance, the Lafrentz caecilian *Dermophis oaxacae*, a blue-black species from Oaxaca, Mexico, and known to the locals as *metlapil*, is fervently but mistakenly believed by them to be very venomous and willing to bite.

Clearly, therefore, there are persuasive precedents on record for taking seriously the possibility that the death worm is more sinned against than sinner, with its lethal powers owing more to human imagination than physiological evolution.

THE BIGGEST AMPHISBAENID IN THE WORLD?

The squamatans comprise a taxonomic order of scaly reptiles traditionally divided into three sub-orders. Two are very familiar—the lizards (Sauria=Lacertilia), and the snakes (Serpentes=Ophidia). The third, conversely, is virtually unknown except to zoologists, who refer to it as *Amphisbaenia,* and to its 140-odd species as amphisbaenids or worm-lizards.

Like caecilians, these slender, limbless, burrowing vertebrates are deceptively similar in outward form to earthworms. Many are red or brown in color (though a few are handsomely spotted in black and white), and come complete once again with heavily ringed bodies terminating in a near-identical, virtually featureless head and tail.

An amphisbaenid's head is blunt and its skull heavily reinforced with bone to facilitate an exclusively fossorial existence—an existence having little use for eyes, which are therefore tiny and covered by a layer of scales. The tail is also blunt, and is waved in the air if the animal is disturbed while at the soil surface—fooling potential predators into attacking its tail (which in some species can be shed), instead of its head (which can't!). Indeed, human observers often mistakenly believe that these reptiles have a head at each end of their body, like the legendary vermiform monster after which they are named.

Amphisbaenids feed upon soil-dwelling arthropods and earthworms, which they encounter while tunnelling or when emerging above-ground, either at night or if flooded out of their burrows during rainfall. They will even lie in wait for prey victims, just below the soil surface, because they are very sensitive to vibrations. Once they detect any movement overhead from unwary prey, they lunge up through the soil and seize it.

Amphisbaenids can be found in tropical or subtropical regions of southern Europe, Africa, and the Middle East, and both American continents, including arid regions, where their scaly skin is better able to resist water loss than the smooth, moist skin of earthworms and caecilians. Is it conceivable, therefore, that the *allghoi khorkhoi* is an undiscovered species of giant, desert-adapted amphisbaenid?

This identity is favored by Czech cryptozoologist Jaroslav Mareš, who rightly emphasizes the external similarity of amphisbaenids to the nomads' description of the death worm. Moreover, the latter's locomotion, which has been likened by nomads to that of a worm, can also be reconciled with an amphisbaenid identity. Their mode of progression through the soil was well described by herpetologist Chris Mattison in *Lizards of the World* (1989):

> The animal moves along its tunnel by sliding its rings of skin forward, over the body itself, bracing them against the walls of the tunnel and then pulling the body forward by means of muscles connecting the body wall to the inside surface of the skin.

This distinctive mode of progression is aptly termed concertina or rectilinear locomotion, and is also exhibited by certain species of snake. Amphisbaenids are equally worm-like on the surface too, hitching their body along in a straight line (and sometimes even looping vertically like a giant caterpillar), instead of undulating laterally like typical snakes and legless lizards. In the eyes of a layman, therefore, an amphisbaenid would certainly bear more of a resemblance to a worm than to most snakes or lizards—and a giant amphisbaenid would thus resemble a giant worm, which is how the nomads describe the death worm.

Conversely, although a few species of amphisbaenid can exceed two feet in length, none attains the size attributed to the death worm. Nor does any squirt fluid from its body in the manner described for this bewildering creature. In any case, whereas one assumes that if the death worm really does eject fluid (toxic or otherwise), it does so from its mouth; amphisbaenids tend to raise their tails, not their heads, toward potential predators. So unless the *all-*

ghoi khorkhoi yields death via deadly defecation, this poses another problem in reconciling Mongolia's most feared mystery beast with an amphisbaenid identity. Its zoogeography does too, as the only known Asian amphisbaenids are confined to Asia Minor (the Middle East).

Moreover, none of the amphisbaenids currently recorded by science is venomous. However, as with certain caecilians, this has not prevented native superstition falsely bestowing such properties upon various species of amphisbaenid. One of North Africa's representatives, for example, the sharp-tailed worm-lizard *Trogonophis wiegmanni,* is feared as an exceedingly poisonous creature by the local people, because they erroneously believe it to be the juvenile form of *Cerastes cornutus*—the genuinely venomous horned viper! Hence it is by no means impossible that if the death worm is an undescribed species of giant amphisbaenid, its venom-spitting ability is nothing more than a fanciful invention.

WINGED SALAMI AND WORMS WITH EARS— ADJUDICATING WITH AJOLOTES

Earlier, I noted the bizarre claim by a Mongolian shepherd that in the Chongor Gobi there is a death worm of sorts that resembles salami but has a pair of wings at its rear end!

In reality, however, the shepherd's claim is not quite so weird as it might initially seem. Is it possible that this "winged salami" is actually a worm-like reptile with only a single pair of legs, which, in a desert terrain, would probably be splayed or spatulate in shape for digging, and might therefore resemble wings? As it happens, three species of just such an animal are already recorded by science, and known as as ajolotes. Moreover, these amazing creatures *(Bipes* spp.) are actually amphisbaenids—the only amphisbaenids with legs.

Sporting long, exceedingly worm-like bodies but less than two feet in total length, all three ajolotes are endemic to the arid terrain of western Mexico (including Baja California). Each species has a single pair of small but well-formed front limbs with broad feet, positioned just behind their head. Indeed, these are often mistaken for ears by observers.

AJOLOTE

Morphologically, an ajolote would provide a very plausible explanation for the "legged death worm" of Chongor Gobi. Or at least it would if we are willing to assume that, because amphisbaenids' heads and tails are very similar in appearance—and because he had only ever seen dead specimens of his "salami" mystery beast—the shepherd may have mistakenly thought that his mystery beast's "wings" (i.e. its spade-shaped legs) were at its body's rear end instead of its front end. Even so, as ajolotes constitute an exclusively New World taxonomic family, this identification is zoogeographically unsatisfactory.

Of course, it is conceivable that an Old World amphisbaenid lineage could have given rise to an ajolote counterpart by convergent evolution, but aside from the account of the Chongor Gobi mystery beast (and also Boldoos's all-too-briefly spied beast?), there is not a scrap of evidence to suggest this. Alternatively, the latter creature could be a species of two-legged lizard—of which a number of species are already known to science.

Yet regardless of its morphological similarity or otherwise to species already known to science, it is still unclear whether or not this legged mystery beast from the Chongor Gobi is indeed one and the same as the seemingly limbless death worm reported elsewhere in the Gobi. Bearing in mind, however, that nomads prefer to keep as great a distance as feasible between themselves and any death worm that they encounter, it is not impossible that the *allghoi khorkhoi* really does possess a small, inconspicuous pair of limbs—but which can only be discerned if viewed at extremely close range or in dead specimens, and whose presence is therefore not widely realized, even among the nomads.

WIDE-EYED AND LEGLESS? LIZARDS SANS LIMBS

Several taxonomic families of true lizards, i.e. lacertilians (saurians), contain legless species. The most famous are various limbless anguids (including the slow worm *Anguis fragilis* and the European scheltopusik *Ophisaurus apodus*), and the Australasian snake-lizards or pygopodids (they lack front limbs, and their hind limbs are reduced to scarcely visible scaly flaps). Even so, these bear little resemblance to the death worm, as they have well-differentiated heads with large, wide, readily perceived eyes and long, pointed tails.

Much more vermiform are certain tiny-eyed legless skinks, such as the small, burrowing *Acontias* skinks indigenous to sandy terrain in South Africa and Madagascar, and the larger, forest-dwelling *Feylinia cussori* from tropical Africa. There are also some skinks with hind limbs but without front limbs, such as Australia's *Lerista bipes* and *L. labialis*. Two small families of limbless worm-like lizards are the blind lizards (dibamids) of Mexico, southeastern Asia, and New Guinea, whose females are legless, but whose males possess a pair of vestigial hind legs in the form of flipper-like stumps; and the anniellids, comprising two species of tiny, shiny-bodied species native to California and Mexico's Baja California.

Evidently, limblessness and semi-limblessness are not uncommon phenomena among lizards, having evolved independently on several occasions within this reptilian sub-order. Having said that, all of the points raised for and against the death worm being an amphisbaenid also apply in relation to its candidature as a legless lizard—except, that is, for its worm-like mode of locomotion. For whereas this is indeed compatible with an amphisbaenid identity, legless lizards move via horizontal undulations, like snakes.

All known amphisbaenids are non-venomous, but there are two species of venomous lizard—the Gila monster *Heloderma suspectum* from the deserts of the southern U.S. and northwestern Mexico, and the beaded lizard *H. horridum* of Mexico and Guatemala. Both species, however, have two pairs of well-formed limbs, and release their venom only when biting their prey, not by spraying or squirting it like the death worm.

One very curious facet of the death worm's reported behavior is its propensity for emerging onto the surface of the sand only during the Gobi year's two hottest months, June and July. It is well known that as reptiles are poikilothermic ("cold-blooded"), their body temperature varies with that of their surroundings. Accordingly, in temperate climates reptiles sunbathe during sunny days to attain an optimal body temperature for energetic activity, and enter a period of hibernation or inactivity during cold conditions.

Faced with intense heat, however, many reptiles tend to seek shelter in cooler, shady locations (including burrows) during the daytime—especially during the summer months—in order to avoid over-heating and eventual death. Such reptiles as these, which include most desert-dwelling lizards, are known as thermal regulators, actively regulating their body temperature by way of their behavioral activity. The obverse side of this physiological coin is represented among desert reptiles by the thermal conformers, whose preferred habitat acts as a buffer against environmental temperature extremes, so that they rarely need to move from one locality to another for temperature-related reasons.

Prime examples from this latter category of reptiles are desert-dwelling amphisbaenids and legless lizards, which generally remain underground, their body temperature maintained at the same temperature as that of the surrounding sand. Consequently, if the death worm is indeed a fossorial reptile belonging either to the lizard or to the amphisbaenid sub-order, there does not seem to be any obvious thermal reason why it should emerge onto the surface during the hot summer months. On the contrary, if it were to emerge at all, I would have expected it to have done so during the colder months of the year—when remaining permanently beneath the surface may not have yielded sufficient warmth in order for it to maintain its body at its optimal functioning temperature.

SERPENTS IN THE SAND—FROM DEATH WORM TO DEATH ADDER?

Just like the amphisbaenids and lizards, there are a number of specialized desert-dwelling snakes that are sufficiently vermiform to conjure up favorable comparisons with the traditional death worm description. One taxonomic family that readily comes to mind, for example, comprises the blind snakes (e.g. *Typhlops* spp.). These are small, cylindrical serpents with glossy brown scales, scale-covered eyes, and a spine-tipped tail. Some do inhabit arid, sandy regions, and are rarely seen above the surface. There are species in much of Africa, southern Asia, Australasia, Central America, and the upper half of South America. Another family of worm-like burrowing species are the thread snakes, distributed throughout Africa, most of Central and South America, and parts of western Asia. They have tiny eyes and mouth, a blunt snout and tail, and can be easily mistaken for large earthworms. Like the blind snakes, however, they are inoffensive and non-poisonous.

When I spoke about the death worm to Mark O'Shea, one of Britain's leading field herpetologists and snake-handlers, he opined that it could be a species of *Eryx* sand boa. Native to arid zones in western Asia, Africa, and southeastern Europe, these burrowing, thick-set snakes with cylindrical bodies, blunt heads, and similarly blunt tails certainly recall the death worm in superficial morphology, and spend much of their time partially or entirely buried in the sand, lying in wait for unsuspecting lizards and rodents to approach near enough to be seized. Unlike the death worm, however, sand boas rarely come out onto the surface of the sand during the day, and they possess distinct eyes.

In addition, the death worm can be readily spied due to its striking dark red color—could this be deliberate warning coloration, signaling the worm's lethal prowess to would-be attackers? Conversely, sand boas display cryptic coloration, for camouflage purposes, preventing them from being detected by their prey when lurking in the sand.

Moreover, these snakes rarely exceed three feet in length, and are non-poisonous. Mark suggested that the death worm's alleged ability to squirt venom may well be folklore, which is undeniably a possibility, as noted for all of the previous taxa of animals considered here as putative identities for the death worm. In contrast, the morphological and behavioral differences between the death worm and the sand boas cannot be so readily discounted.

A very different type of desert snake that is indeed dark red sometimes is South Africa's horned adder *Bitis caudalis*—just one of several sand-inhabiting members of the viper family recorded from the world's deserts. Unlike sand boas, these species are venomous (though they release their venom when biting their prey, not by squirting it into the air), and live on or just below the surface of the sand, rather than indulging in a predominantly subterranean lifestyle. Unfortunately for those attempting to reconcile such species with the death worm, however, these snakes tend to possess readily visible, heavily keeled scales, large eyes, and the ability to flatten themselves by spreading their ribs, so that they may vanish rapidly beneath the surface of the sand when detecting the approach of a potential prey victim. No such behavior, conversely, has apparently been documented for the death worm.

Nevertheless, there remains one excellent contender among the serpent sub-order for the veiled identity of the death worm. This contender was first brought to my attention when discussing the death worm with Prof. John L. Cloudsley-Thompson—a world-renowned expert on the biology of deserts and editor of the *Journal of Arid Environments*.

In a letter to me from October 1996, Prof. Cloudsley-Thompson offered the following pertinent thoughts regarding the death worm:

> I would have thought that the animal must be a snake...I do not think that any of the vipers spray poison as cobras do but, like the puff adder, they can reach a large size. One idea that occurs to me is that the Australian 'death adder' [*Acanthophis antarcticus*] is really an elapid snake which has evolved to occupy a viperine ecological niche. Could the 'death worm' be an unknown species that has done the same? The king cobra is not only one of the largest of venomous snakes but, also, one of the most deadly...

> Many poisonous snakes will spray drops of venom for a short distance, and three species of cobras—the ringhals, black-necked, and (some) Indian—have developed the technique of aiming at the eyes of an aggressor [these are the famous spitting cobras]. The twin jets, according to Bellairs, can be sprayed >2 m [six feet]. Cobra venom is rapidly absorbed from the eye (and perhaps by delicate skin?) but not the venom of vipers. This again might suggest a snake like *Acanthophis* spp. spraying an extremely toxic and rapidly acting venom....

Death adders have broad heads, *narrow* necks, stout bodies and thin *rat-like tails* ending in a curved soft spine (? = pointed at both ends)–unlike other snakes. They are extremely toxic. They occur in Australia, New Guinea and islands west as far as Ceram (=Seram), according to Harold Cogger. As they occur in the East Indies, could not a species also be found in East Asia too?...

Clearly, it is a matter of weighing one thing against another, but my guess would be a new species or genus of death adder. If this idea is of use I would be very happy to make a present of it to you, to acknowledge or not—as would be best for your publication. No doubt the animal's food would be chiefly small rodents and, perhaps, other reptiles. It is not clear why some snakes (e.g. true coral snakes) are so much more venomous than would appear necessary.

Prof. Cloudsley-Thompson's idea that the death worm may in reality be a hitherto-unknown species of death adder is quite fascinating, and is the first identity that offers a factual explanation for the nomads' reports of this cryptid's venom-spraying talent.

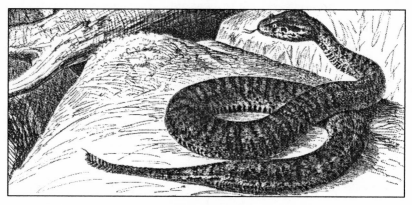

DEATH ADDER

THE BEASTS THAT HIDE FROM MAN

The elapids are a taxonomic family of snakes housing such familiar if fatal species as the cobras, mambas, kraits, and also the death adder. As outlined above, this unusual snake has acquired via convergent (parallel) evolution a remarkably viper-like (i.e. adder-like) appearance, thus occupying the ecological niche left vacant in Australia by the true vipers—as there is no viper species native to this vast island continent. In other words, the death adder is an elapid impersonating a viper. But what would happen if a death adder also evolved the venom-spitting ability exhibited by certain of its fellow elapids, i.e. the spitting cobras?

As indicated by Cloudsley-Thompson, the result could be a specialized snake able to spray deadly fast-acting, skin-absorbing venom at any would-be attacker from a not inconsiderable distance—thereby corresponding with the descriptions given by the Gobi nomads in relation to the death worm, even including those ostensibly bizarre claims that the worm can kill from a distance, without making direct contact with its victim. In addition, if absorbed through the eyes, the venom could work its lethal effect without needing to be corrosive—although the nomads' testimony certainly suggests that the venom is acidic.

One important matter needing to be discussed here, however, is the precise shape of the death worm's anterior and posterior body regions. Some reports seem to imply that the ends of its body are blunt or truncated. In marked contrast, however, is the eyewitness report quoted earlier in this chapter, in which she claimed that the death worm was "bound into points at both ends." This suggests that it was not blunt but pointed at each end (as depicted in Philippa Foster's picture)—and, if so, this description does accord well with the narrow-necked, spine-tailed morphology of the death adder.

One notable problem with a serpent identity for the death worm, at least on first reflection, is Ivan Mackerle's assertion that it moves in a worm-like manner by contracting and expanding its body. Snakes, conversely, typically move via horizontal undulations, but as I noted when discussing the amphisbaenids, some snakes can perform concertina (rectilinear) locomotion. Moreover, in a separate section of his letter to me of October 1996, Prof. Cloudsley-Thompson also offered this as an explanation for the apparent locomotory contradiction between death worm and snake:

> [Having suggested a species of death adder as a possible identity] this leaves the problem of expansion and contraction of the body. Could this not be explained in terms of rectilinear movement? (Angus Bellairs called it "concertina" movement, which does suggest contraction and expansion of the body.) In every case of this, waves of muscular contraction pass posteriorly over the muscular under surface of the body while the animal is in motion. Apparently many snakes can move in this way although it is most noticeable in boas and vipers. Rectilinear locomotion is used chiefly when stalking prey or moving through narrow spaces.

Here, then, is a very satisfactory reconciliation between the allegedly worm-like movements of the death worm and an ophidian identity.

Another aspect of the death worm's movements that can be linked with snake locomotion is its alleged "sideways sweeping about." It is well known that certain desert snakes, notably the American sidewinder rattlesnake *Crotalus cerastes* and the Namibian sidewinder viper *Vipera peringuey*, move across the sand via a peculiar laterally directed mode of locomotion known as sidewinding. This serves two purposes. It enables the snake to move readily across a highly unstable substratum that would otherwise pose great problems for a limbless animal. And as sidewinding involves raising portions of its body up off the surface, the snake is successfully limiting its body's extent of contact with the hot sand. Judging from the nomads' description of its movements, perhaps the death worm performs a type of sidewinding too? Ivan Mackerle has noted that it leaves a "wavy trail" when moving on the surface of the sand—reminiscent of the J-shaped traces left by sidewinding snakes?

As for temperature conditions: all that I have already discussed in relation to amphisbaenids and lizards applies to desert snakes too. So not even a snake identity can explain why the death worm should only emerge during the two hottest Gobi months.

LAST OF ALL—AND LEAST OF ALL?

Three identities remain to be considered here, as they have been proposed in the past. In my opinion, however, they are all highly unlikely, to say the very least.

Ivan Mackerle has informed me that Dr. Namhajdorz, a Mongolian zoologist, considers that the death worm may actually be an unknown species of large poisonous insect, allied to the European bombardier beetles *Brachinus* spp. (belonging to the carabid family of ground beetles), or an undescribed species of antlion. Every species of bombardier beetle can emit a fine spray of noxious fluid from its abdomen's tip as a defense mechanism, precisely directing the spray at an attacker in order to startle it into releasing the beetle. This, no doubt, is what inspired Namhajdorz to consider these insects as a death worm identity.

Personally, however, I fail to see how even a uniquely large bombardier beetle (those currently documented by science do not usually exceed one quarter of an inch!) can possibly be likened to a giant worm-like animal, bearing in mind that such beetles have a readily differentiated head, thorax, and abdomen, as well as a pair of large compound eyes, long antennae, a pair of wings, a pair of wing cases, and three pairs of legs!

True, the larva of carabids (including bombardier beetles), known as a campodeiform larva, is long and somewhat worm-like, lacking wings and compound eyes, but it still possess three pairs of legs, and is even smaller than the adult beetle. And it goes without saying, surely, that if, somehow, there could be an undiscovered species of carabid beetle in the Gobi that produces larvae up to five feet long, the adults into which these horrors later transform would truly be monstrous (not to mention physiologically impossible!) in every sense.

As for an unknown species of ant-lion: these insects are related to the familiar lacewings, and superficially resemble weak-flying dragonflies in the adult form. The larval ant-lion, conversely, digs a conical pit in sand, and lurks unseen at the bottom, waiting for hapless ants or other small insects to fall into the pit and slide down to the bottom—where this predatory juvenile is waiting to seize its victims in a pair of massive jaws. Once again, however, such creatures bear no resemblance to the local descriptions given for the death worm, and certainly could not attain the size claimed for it.

The third equally unlikely identity offered up for serious consideration is that the death worm could be an unknown species of giant mole. This prospect was duly documented and favored by B. Avirmed in his various death worm publications, in which he included the account of the driver B. Boldoos, who encountered an odd burrowing mystery beast with rear limb-like projections in Eezh Chairchan, western Mongolia. However, all of the many "true" (i.e. traditional) death worm reports on file are evidently describing a totally different type of animal—one with a long limbless body, and lacking a visibly differentiated head or tail. Thus it is clear that Boldoos' beast has nothing to do with the "true" death worm, and need not be considered further in this present account.

THE BODY ELECTRIC?

So far, the death worm has been shown not to be dramatically different from various other anguinine species inhabiting deserts. As we have seen, even its alleged venom-spraying capacity may have a foundation in fact rather than belonging exclusively to the realms of local folklore. But what about its most amazing purported ability? Is it conceivable that the death worm can actually kill by electrocution? And if not, how can we explain the many reports claiming that it can kill instantly by touch—even by indirect touch, as when prodded with a metal rod—and blaming it for the instantaneous deaths of entire herds of camels?

In order to electrocute anything, the death worm would need to possess specialized electric organs for generating electricity, but the evolution of such organs is not, in itself, particularly implausible. After all, electric organs, developed from modified muscle, are already known to occur in six totally separate taxonomic groups of fish.

These are: the electric catfish, electric rays, electric stargazers, gymnotids (plus their famous cousin, the electric eel), mormyrids (elephant trunk fishes), and rajid skates. Moreover, as these groups are all wholly unrelated to one another, and as in each group the electric organs are derived by a modification of a different type of muscle, the evolution of these organs has clearly occurred six times, completely independently.

This in turn favors the possibility of such organs evolving again, a seventh time—but surely not in a beast so radically different from any fish as the death worm? Yet as my research into the nature and function of electric organs reveals, this cannot be ruled out. Quoting from an earlier death worm article of mine (*Strange Magazine*, fall 1995):

> Electric organs are composed of flattened cells called electroplaques, stacked in vertical columns. Each electroplaque normally produces only about 0.1 volt, but they are generally connected to one another in series, whereas the columns are connected to one another in parallel. Freshwater conducts electricity less efficiently than seawater—hence to generate the higher voltages required for overcoming this electrical

resistance, freshwater electric fishes usually have fewer but taller columns of electroplaques than marine electric fishes (which tend to generate lower voltages but higher currents). The electrical discharge will occur either via the fish's own actions or via an external stimulus (such as something touching the fish).

Both the African mormyrids and the South American gymnotids (knifefishes) are freshwater groups, but they do not emit high voltages, because their electric organs create an electrical "force field" merely for navigation and detecting the approach of potential prey.

This is also the function of two of the three electric organs possessed by the gymnotids' formidable relative, the freshwater electric eel *Electrophorus electricus* of Amazonia—attaining a total length of up to 10 ft (most of which is taken up by its electrical apparatus). However, the third, and largest, organ, whose positive pole is located towards the eel's head (from where the discharge originates), contains approximately 120 electroplaque columns, each consisting of 6,000 to 10,000 electroplaques.

The result of this fearsome battery is a modest current of 1 amp, but with a discharge of up to 550 volts—released into the eel's watery medium to stun any frogs and fishes (yet powerful enough to kill a horse or human) detected by its "force field".

If the death worm possesses a similar physiological arrangement, it would be equally devastating, but because sand is even less effective at conducting electricity than freshwater, a victim would generally need to make direct physical contact with the creature, or via an efficient conducting substance, like the geologist's iron rod, to receive an electric shock. Furthermore, the worm would need to generate considerably more current than the electric eel's 1 amp version in order for such a shock to be lethal to humans.

Can such a scenario also shed light on the nomads' claims that the death worm is even able to kill without making any physical contact with its victim? Once again, my research suggests that this may indeed be possible:

However, if the death worm (presumably acting as a capacitor, and generating electricity as it moves beneath the sand) should find itself separated by only a very short distance from physical contact with a living creature capable of conducting electricity, it is possible that the worm could transmit a high-voltage electrical discharge directly across the small intervening distance, rather like an animate spark plug—thereby stunning or even killing the creature without actually touching it. This would yield a factual basis for the Mongolians' belief in such an ability, with more dramatic claims (i.e. that it can kill from a distance of several feet) deriving from fear-engendered exaggeration.

If the death worm can truly generate and discharge electricity in this manner, it is unquestionably unique among terrestrial fauna, but not unthinkably so. My own feeling regarding its alleged electrocution ability is that if it were solely the product of native superstition, then surely at least some of the great many desert-inhabiting vermiform beasts already described by science (and discussed above in this chapter) would have inspired similar fallacies among their human neighbors. Yet none of them appears to have done so.

The evolution of such an extraordinary physiological process by just a single species of earthworm, amphisbaenid, legless lizard, or snake (or whatever else the death worm might be) may well seem unlikely—but it is not unprecedented. More than 2,000 species of catfish are presently known to science. However, there is only one electricity-generating species, *Malapterurus electricus*. Moreover, it is sufficiently distinct from all other catfish species to be set apart by ichthyologists in a taxonomic family all to itself.

In 1994, Drs. Theodore Vonstille and W.T. Stille from the Envi-Sci Centre of Florida's Winter Park proffered a very novel extra purpose for the rattlesnakes' rattle—proposing that it generates electrostatic charge that can be utilized in detecting prey and hiding places. They revealed that a motionless rattle has no charge; but when it is rapidly vibrated, positive charges of 50 to 100 volts are produced, indicating that a vibrating rattle generates electrostatic charge. Leading on from this remarkable finding, is it conceivable, therefore, that the death worm may generate electricity externally, rather than via internal electric organs?

Michel Raynal has suggested that perhaps the death worm gains an electrical charge via the phenomenon of triboelectricity (frictional electricity), i.e. generating electricity by the rubbing of its body against the enveloping sand as it burrows beneath the surface. Although an intriguing theory, even if it is true I very much doubt that any charge so engendered would be potent enough to kill something as large as a human or camel—though it may be noticeable enough to have been incorporated (and exaggerated accordingly) within the nomads' corpus of traditional legends and lore.

Yet another thought-provoking explanation for the death worm's supposed ability to kill without making physical contact with its victims is that it can somehow emit inaudible but lethal ultrasonic pulses (i.e. pulses of sound with a frequency exceeding 20,000 Hz). This intriguing notion was briefly considered by Jaroslav Mareš in an article from 1996, but he was unable to proffer a satisfactory mechanism via which the worm could achieve such an effect. In any event, death by ultrasonic bombardment would be far less specific in action than electrocution, killing everyone within range, rather than just a single person.

Consequently, not only can this not offer an explanation for the death of the geologist who prodded an *allghoi khorkhoi* with the iron rod, but also it cannot explain how his colleagues, who were all nearby at the time, survived.

IN CONCLUSION

As mentioned earlier, attempts to decide upon a plausible identity for the death worm, if it is indeed a bona fide mystery beast, are irrevocably influenced (in the present absence of tangible physical evidence) by the proportion of this creature that owes its origin not to biology but rather to local folklore.

Assuming that its death-dealing capabilities are fictitious, I consider the most reasonable identity for the *allghoi khorkhoi* to be an unknown species of giant amphisbaenid (or—less plausibly, but even more dramatically if correct—a radically novel species of giant water-impervious oligochaete), with its predilection for sunbathing on the surface of the sand during the Gobi's hottest months just another example of unfounded local folklore. Perhaps it also surfaces during other months, but in areas less readily traversed by nomads, and hence is less commonly spied during those months.

If, conversely, it can indeed squirt or similarly emit a lethal venom, a currently undescribed species of specialized snake, most probably allied to the death adder or some other elapid, is certainly the most plausible identity available.

And if it can kill by electrocution? This talent would distance it so emphatically from all previously documented desert fauna that, without having first examined a specimen, I would hesitate to offer any taxonomic categorization whatsoever for such an astonishing animal.

Cryptozoological skeptics may seek to dismiss the death worm as an example of "foaflore" (i.e. a typical "friend of a friend" tale, lacking any traceable origin or firsthand testimony), due to the frequent reluctance of the nomads to speak of it (or even mention it by name) and to admit having spied it personally. Moreover, while adamant that this creature does exist, they will sometimes deny that it exists in their particular region, and will send European investigators elsewhere to look for it.

However, such evasive behavior on the part of local people in relation to a given animal species is by no means unique. There are many species already known to science, including various lemurs in Madagascar and even the tiger in certain parts of India (e.g. Purnia), that are treated in precisely this manner by the native peoples sharing their domain. Induced by superstitious fear that is often (but not always) groundless, such reluctance to talk about these animals certainly assists in keeping them hidden from scientific scrutiny, and may well explain why certain cryptids, including the death worm, continue to elude formal discovery.

Many cryptozoological analyses can incorporate clues drawn not only from the fauna of the present, but also from that of the past. In the case of the death worm, however, nothing remotely like it has ever been documented in the paleontological literature. Of course, if it is an invertebrate worm, this is not too surprising, because soft-bodied animals rarely yield fossils anyway. Equally, impressions of soft organs capable of performing the deadly deeds attributed to this creature would probably not be preserved even in vertebrate fossils.

Clearly, then, we must look to the future, and not to the past, if we are ever to obtain a conclusive answer to the unique riddle posed by the Mongolian death worm. If such a creature truly exists, its discovery and formal recognition by science will unquestionably be one of the most sensational zoological events of modern times. However, the Gobi is an exceedingly big place, as well as an inaccessible and highly inhospitable one, so it is likely that its most mysterious inhabitant will remain mysterious for a long time to come.

And who knows, perhaps that may be no bad thing. In the words of Ben Okri, from *Astonishing the Gods* (1995):

> When you make sense of something, it tends to disappear. It is only mystery which keeps things alive.

CHAPTER 3

Raven and the Terror Bird!

*Ghastly, grim, and ancient Raven, wandering
from the nightly shore,—
Tell me what thy lordly name is on the night's
Plutonian shore?*

EDGAR ALLAN POE—*THE RAVEN*

A MAJOR FIGURE IN AMERINDIAN MYTHOLOGY ACROSS much of North America is Raven. Specific details concerning its appearance vary from one tribal group to another, but the traditional description quoted by certain native tribes of the southern U.S. is of particular cryptozoological interest. This is because it tantalizingly recalls an incredible bird hitherto assumed to have died out long before humans first entered its North American domain, and so monstrous in form and behavior that paleontologists informally call it a "terror bird."

NORTH AMERICA'S FEATHERED TITAN

One of Africa's strangest birds is a long-legged vaguely stork-like species known as the secretary bird *Sagittarius serpentarius*. Standing three to four and a half feet tall and famous for its snake-killing talents, it has long posed a problem for zoologists anxious to determine its precise taxonomic relationship to other species of bird. Traditionally, it has been classed as a bird of prey, but some researchers claim that it is most closely related to a couple of intriguing South American species known as seriemas. These are certainly reminiscent of the secretary bird, but are also of great paleontological interest, because they appear to be the closest liv-

NORTH AMERICAN INDIAN SHAMAN WEARING COSTUME OF BIRD DEITY

ing relatives of a group of spectacular extinct flightless birds known technically as phorusrhacids, but referred to colloquially (and very aptly) by their researchers as terror birds.

Plenty of phorusrhacid fossils have been found in South America, which for many millions of years had been an island continent, just like Australia still is, until it finally became linked to North America via the emergence from the sea of the interconnecting isthmus of Panama, around two to four million years ago. Prior to this event, however, the ecological niche of "big carnivore," occupied in many other regions of the world by dogs, cats, and other placental mammals, had been filled to a large extent in South America by the phorusrhacids.

The first known phorusrhacids were modest-sized species, but by the mid-Miocene (around 15 million years ago), the plains of Patagonia were being stalked by the terrifying spectacle of *Phorusrhacus inflatus*. Standing five to six feet tall, this flightless but very efficient running bird possessed a massive hooked beak, bore four huge, razor-sharp talons on each foot, and preyed upon ungulates and other very sizeable herbivorous mammals.

As the top predators in the island continent of South America, the phorusrhacids ruled supreme. Once it became connected to North America via the Panamanian isthmus, however, large placental mammalian carnivores such as dogs, cats, and bears soon traveled down into South America, where they established themselves and decimated most of the unique herbivorous mammals that had evolved in isolation there. For a long time, paleontologists assumed that the phorusrhacids had also succumbed, losing out in competition with these invading furry predators from the north. In 1963, however, this assumption was dramatically swept aside by the remarkable news that some fossilized remains from a huge species of phorusrhacid had been found in Florida. Only two million years old, it was clear from these fossils that far from dying out, some phorusrhacids had actually succeeded in journeying up along the Panamanian isthmus from South into North America, and had established themselves in Florida.

The new, North American phorusrhacid was christened *Titanis walleri*, and attracted great attention because some researchers estimated that this flightless, flesh-eating bird had stood 10 to 12 feet tall—more than two feet taller than the ostrich! Truly a terror bird!

More recently, it has been downscaled by certain researchers to a somewhat more modest six feet or so—but a human-sized carnivorous bird would still be a frightening prospect by anyone's standards.

ARMED AND VERY DANGEROUS!

The two most remarkable discoveries concerning *Titanis* did not occur until the 1990s. The first came after paleontologist Dr. Robert Chandler began seeking its fossils in the Santa Fe River passing through Florida. During one search, he found some *Titanis* wing bones. This was very important because phorusrhacid remains had been notable for their absence of wing skeleture, leading researchers to assume that during the course of evolution these birds' wings had degenerated until they had eventually vanished altogether. Chandler, however, proved that this was not so. Moreover, the wing was hardly vestigial, measuring three feet long. But it was its structure that so amazed Chandler, for the bones had evolved in a manner that had literally converted the wing into an arm, equipped with a three-clawed "hand" or "paw." Chandler believes that *Titanis* used its arms to prevent its prey victims from attacking it with their horns or hoofs, and also that it could manipulate its victims with its hands, and impale them with its claws. It would thus behave in a manner unexpectedly reminiscent of certain bipedal carnivorous dinosaurs, such as the infamous *Velociraptor*.

Even more dramatic, however, was a *Titanis* revelation by paleontologist Dr. Jon Baskin. He recently found a *Titanis* toe in a gravel pit along the Nueces River—in Texas. But that was not all. The other fossils found with it in the gravel pit came from two well-delineated time periods. One was five million years ago, i.e. quite some time before South America rejoined North America, enabling the ancestors of *Titanis* to travel north. Hence it would seem that the *Titanis* toe could not possibly be from that period. However, the other time period represented by fossils found with the *Titanis* toe was a mere 15,000 years ago! In other words, if the toe were not five million years old (which our current zoogeographical knowledge dictates that it could not have been), it must be no more than 15,000 years old!

If so, this is highly significant, because instead of dying out entirely around two million years ago (as hitherto assumed), it means that *Titanis* not only had existed in Texas (as well as Florida) but had survived here until at least as recently as 15,000 years ago—by which time, according to current belief, humankind had reached the southern regions of North America. Suddenly, we are faced with the amazing possibility that some humans had actually encountered the most terrifying species of bird ever to live on planet Earth!

RAVEN TO THE RESCUE?

If only there was some other evidence to support this fascinating hypothesis. Perhaps, as lately discussed by Czech cryptozoologist Dr. Jaroslav Mareš in his book *Svet Tajemných Zvířat* (1997), there is. As he points out, the Kwakiutl and Haida tribes of the southern U.S. describe the mythological being Raven as a monstrous giant bird with a massive hooked beak, long sturdy legs on which it runs rapidly across the ground, and a feathered body like other birds—but possessing front paws with claws, instead of wings!

Now where have we come across a bird that possessed clawed paws (or hands), instead of wings, before? Is it just a coincidence that legends of such a bird can be found in the one region of North America where a species fitting this description apparently did coexist with humans? Today, *Titanis* is long since gone, but perhaps its formidable *memory* lingers on.

CHAPTER 4

The Strange Case of Conan Doyle's Brazilian Black Panther

The cat (if one may call so fearful a creature by so homely a name) was not more than ten feet from me. The eyes glimmered like two disks of phosphorus in the darkness. They appalled and yet fascinated me. I could not take my own eyes from them.

SIR ARTHUR CONAN DOYLE—"THE BRAZILIAN CAT"

THIS CASE IS A MYSTERY WELL DESERVING OF SHERLOCK Holmes's attention—due not only to its complexity, but also to the fact that it features Holmes's creator, Sir Arthur Conan Doyle.

A BRAZILIAN MYSTERY CAT—FICTION OR FACT?

Several years ago, I read a short story by Conan Doyle called "The Brazilian Cat," published in 1922, which featured a huge, ferocious, ebony-furred felid that had been captured at the headwaters of the Rio Negro in Brazil. According to the story: "Some people call it a black puma, but really it is not a puma at all." Yet there was no mention of cryptic rosettes, which a melanistic (all-black) jaguar ought to possess, and it was almost 11 feet in total length, thereby eliminating both puma and jaguar from consideration anyway. Hence I simply assumed that Doyle's feline enigma was fictitious, invented exclusively for his story. But following some later cryptozoological investigations of mine, I am no longer quite so sure.

THE BORDERLAND BLACK PANTHER OF COLONEL FAWCETT

To begin with: in *Exploration Fawcett* (1953), the famous lost explorer Lt.-Col. Percy Fawcett briefly referred to a savage "black panther" inhabiting the borderland between Brazil and Bolivia that terrified the local Indians, and it is known that Fawcett and Conan Doyle met one another in London. So perhaps Fawcett spoke about this "black panther" and inspired Doyle to write his story. But even so, it still does not unveil the identity of Fawcett's panther. Black pumas are notoriously rare—only a handful of specimens have been obtained from South and Central America (and none ever confirmed from North America). Conversely, black jaguars are much more common, and with their cryptic rosettes they are certainly reminiscent of (albeit less streamlined than) genuine black panthers, i.e. melanistic leopards. However, the mystery of Brazil's black panthers is far more abstruse than this.

JAGUARETE

UNRAVELLING THE JAGUARETE'S IDENTITY

As long ago as 1648, traveler Georg Marcgrave mentioned a shiny Brazilian black cat, variegated with black spots, which he dubbed the jaguarete. This felid, clearly resembling a typical melanistic jaguar, was later documented from Guyana, too, and was referred to as the black tiger—a name popularly applied in Latin America to black jaguars. So far, so good.

However, when the famous British naturalist Thomas Pennant described and depicted this "black tiger" in 1771, he portrayed it with plain black upperparts and whitish or ashen underparts—whereas true black jaguars are uniformly black all over. Yet black pumas do indeed have paler underparts, and Pennant's French contemporary, Buffon, actually referred to the jaguarete as the *cougar noire* ("black cougar"), indicating a puma identity for it.

The jaguarete was even given its own scientific name, *Felis discolor*, by German naturalist Johann Christian Daniel Schreber, in 1775. Yet as it seemed to be nothing more than a non-existent composite creature—"created" by early European naturalists unfortunately confusing reports of black jaguars with black pumas—the jaguarete eventually vanished from the wildlife books. Even so, its rejection by zoologists as a valid, distinct felid may be somewhat premature. This is because some reports claimed that the jaguarete was much larger than either the jaguar or the puma—a claim lending weight to the prospect that a third, far more mysterious black cat may also have played a part in this much-muddled felid's history.

THE *YANA PUMA*—AN ALL-BLACK ANOMALY FROM PERU

Dr. Peter J. Hocking is a zoologist based at the Natural History Museum of the National Higher University of San Marcos, in Lima, Peru. Aside from his official work, for several years he has been collecting and investigating local Indian reports describing various different types of mysterious, unidentified cat said to inhabit the Peruvian cloudforests. One of these, of great relevance here, has been dubbed by him "the giant black panther."

In an article published by the journal *Cryptozoology* in 1992, Dr. Hocking revealed that this particular Peruvian mystery cat is said to be entirely black, lacking any form of cryptic markings, has large green eyes, and is at least twice as big as the jaguar. Moreover, the Quechua Indians term it the *yana puma* ("black mountain lion"). This account immediately recalls Conan Doyle's story of the immense Brazilian black cat. The *yana puma* is apparently confined to montane forest ranges only rarely visited by humans, at altitudes of around 1,600 to 5,000 feet. If met during the day, when resting, it is generally passive, but at night this mighty cat becomes an active, determined hunter that will track humans to their camps and has sometimes slaughtered an entire party while they slept, by lethally biting their heads.

When discussing the *yana puma* with mammalogists, Hocking has frequently been informed that it is probably nothing more than a large melanistic jaguar. Yet as he pointed out in his article, such animals do not attain the size claimed for this mysterious felid (nor do melanistic pumas)—and the Indians are adamant that it really is quite enormous. Nevertheless, the *yana puma* could still merely be a product of native exaggeration, inspired by real black jaguars (or pumas) but distorted by superstition and fear.

However, a uniformly black, unpatterned felid does not match either a black jaguar or a black puma—yet it does compare well with Conan Doyle's Brazilian cat. Moreover, as some jaguarete accounts spoke of a black cat that was notably larger than normal jaguars and pumas, perhaps the *yana puma* is not limited to Peru, but also occurs in Brazil. Is it conceivable, therefore, that Doyle (via Fawcett or some other explorer contact) had learned of the *yana puma* and had based his story upon it? If so, it would be one of the few cases on file in which a bona fide mystery beast had entered the annals of modern-day fiction before it had even become known to the cryptozoological — let alone the zoological—community!

CHAPTER 5

A BELFRY OF CRYPTO-BATS

For something is amiss or out of place
When mice with wings can wear a human face.
THEODORE ROETHKE—"THE BAT"

FROM THE EARLIEST TIMES, BATS HAVE BEEN VIEWED AS creatures of mystery—as arcane and abstruse as the opaque shadows of Night that animate so many of their kind. It is certainly true that during the past two centuries science has succeeded in brushing aside many cobwebs of longstanding confusion and credulity enveloping these winged wanderers, but there are a host of others still to be dealt with. Indeed, as disclosed in this chapter, there are certain sources of provocative (and scientifically inconvenient?) but little-known data on file that not only question the validity of some fondly treasured tenets of traditional bat biology, but also provide startling evidence for the existence of several dramatic species of bat still awaiting formal zoological discovery.

THE *AHOOL* AND THE *OLITIAU*—
OR, HOW BIG IS A BAT?

One evening in 1927, at around 11:30 p.m., naturalist Dr. Ernst Bartels was in bed inside his thatched house close to the Tjidjenkol River in western Java, and lay listening to the surrounding forest's clamorous orchestra of nocturnal insects. Suddenly, a very different sound came winging to his ears from directly overhead—a loud, clear, melodious cry that seemed to utter "A-hool!" A few moments later, but now from many yards further on, the cry came again—a single "A-hool!" Bartels snatched his torch up and ran out, in the direction of this distinctive sound. Less than 20 seconds later he heard it once more, for the third and last time—a final "A-hool!" that floated back towards him from a considerable distance downstream. As he recalled many years later in a detailed account of this and similar events (*Fate,* July 1966), he was literally transfixed by what he had heard—not because he *didn't* know what it was, but rather because he *did!*

The son of an eminent zoologist, Dr. Bartels had spent much of his childhood in western Java, and counted many of the local Sundanese people there as his close friends. Accordingly, he was privy to many strange legends and secret beliefs that were rarely voiced in the presence of other Westerners. Among these was the ardent native conviction that this region of the island harbored an enormous form of bat. Some of Bartels's Sundanese friends claimed to have spied it on rare occasions, and the descriptions that they gave were impressively consistent. Moreover, as he was later to discover, they also tallied with those given by various Westerners who had reputedly encountered this mysterious beast.

It was said to be the size of a one-year-old child; with gigantic wings spanning 11 to 12 feet; short, dark-grey fur; flattened forearms supporting its wings; large, black eyes; and a monkey-like head, with a flattish, man-like face. It was sometimes seen squatting on the forest floor, at which times its wings were closed, pressed up against its flanks; and, of particular interest, its feet appeared to point backwards. When Bartels questioned eyewitnesses as to its lifestyle and feeding preferences, he learned that it was nocturnal, spending the days concealed in caves located behind or beneath waterfalls, but at night it would skim across rivers in search of large fishes upon which it fed, scooping them from underneath stones on the beds of the rivers. At one time, Bartels had suggested that perhaps the creature was not a bat but some type of bird, possibly a very large owl, but these opinions were greeted with great indignation and passionate denials by his friends, who assured him in no uncertain terms that they were well able to distinguish a bat from a bird! And as some were very experienced, famous hunters, he had little doubt concerning their claims on this score.

Even so, the notion of a child-sized bat with a 12-foot wingspan seemed so outrageous that he still had great difficulty in convincing himself that there might be something more to it than native mythology and imagination—until, that is, the fateful evening arrived when he heard that unforgettable, thrice-emitted cry, because one of the features concerning the giant bat that all of his friends had stressed was that when flying over rivers in search of fish this winged mystery beast sometimes gives voice to a penetrating, unmistakable cry, one that can be best rendered as "A-hool! A-hool! A-hool!"

Indeed, the creature itself is referred to by the natives as the *ahool*, on account of its readily recognizable call—totally unlike that of any other form of animal in Java, as Bartels himself was well aware.

Transformed thereafter from an *ahool* skeptic to a first-hand *ahool* "earwitness," Bartels set about collecting details of other *ahool* encounters for documentation, and eventually news of his endeavors reached veteran cryptozoologist Ivan T. Sanderson, who became co-author of Bartels's above-cited *Fate* article.

The *ahool* was of special interest to Sanderson, because he too had met with such a creature—but not in Java. Instead, he had been in the company of fellow naturalist Gerald Russell in the Assumbo Mountains of Cameroon, in western Africa, collecting zoological specimens during the Percy Sladen Expedition of 1932. As Sanderson recorded in his book *Animal Treasure* (1937), he and Russell had been wading down a river one evening in search of tortoises to add to their collection when, without any warning, a jet-black creature with gigantic wings and a flattened, monkey-like face flew directly toward him, its lower jaw hanging down and revealing itself to be unnervingly well-stocked with very large white teeth. Sanderson hastily ducked down into the water as this terrifying apparition skimmed overhead, then he and Russell fired several shots at it as it soared back into view, but the creature apparently escaped unscathed, wheeling swiftly out of range as its huge wings cut through the still air with a loud hissing sound. Within a few moments, their menacing visitor had been engulfed by the all-encompassing shadows of the night, and did not return.

After comparing notes, Sanderson and Russell discovered that they had both estimated the creature's wingspan to be at least 12 feet (matching that of the *ahool*). When they informed the local hunters back at their camp of their experience, the hunters were petrified with fear—staying only as long as it took them to shriek *"olitiau!"* before dashing away en masse in the direction furthest from the scene of the two naturalists' encounter! As Dr. Bernard Heuvelmans commented in his own coverage of this incident, within his book *Les Derniers Dragons d'Afrique* (1978), *"olitiau"* may refer to devils and demons in general rather than specifically to the beast spied by Sanderson and Russell, but as these etymological issues have yet to be fully resolved it currently serves as this unidentified creature's vernacular name.

Several authorities have boldly attempted to equate the *olitiau* with a pterodactyl, in preference to a giant bat. Interestingly, in parts of Zimbabwe and Zambia a creature supposedly exists whose description vividly recalls those long-extinct flying reptilians. It is known to the natives as the *kongamato,* and has been likened by them to a small crocodile with featherless, bat-like wings—a novel but apt description of a pterodactyl!

In the case of the *olitiau,* however, Sanderson remained adamant that it was most definitely a bat of some kind, albeit one of immense dimensions—which is why he was so interested by Bartels's account of the *ahool,* which indicated that undiscovered giant bats were by no means limited to Africa. In fact, it is now known that similar animals have also been reported from such disparate areas of the world as Vietnam, Samoa, and Madagascar. Madagascar's version is known as the *fangalabolo* ("hair-snatcher"), named after its alleged tendency to dive upon unsuspecting humans and tear their hair (a belief reminiscent of the Western superstition that bats can become entangled in a person's hair).

If giant bats do exist, how much bigger are they than known

species, and to which type(s) of bat could they be related? The answer to the first of these questions is simple, if startling. The largest living species of bat whose existence is formally recognized by science is the Bismark flying fox *Pteropus neohibernicus,* a fruit bat native to New Guinea and the Bismark Archipelago; it has a total wingspan of five and a half feet to six feet. Assuming that eyewitness estimates for the wingspan of the *ahool* and the *olitiau* are accurate, then both of these latter creatures (if one day discovered) would double this record. The question of their taxonomic identity is a rather more involved issue.

Known scientifically as chiropterans ("hand-wings" because the greater portion of their wings are membranes of skin extending over their enormously elongated fingers), the bats are traditionally split into two fundamental but quite dissimilar suborders.

The megachiropterans, or mega-bats for short, comprise the fruit bats. Often called flying foxes as many have distinctively vulpine (or even lemur-like) faces, these are a mostly large, primarily fruit-eating species, and rely predominantly upon well-developed eyes for avoiding obstacles during night flying.

The microchiropterans, or micro-bats for short, comprise all of the other modern-day bats. Often small and insectivorous, but also including some vegetarian and fish-eating forms, as well as the notorious blood-sipping vampires of tropical America (certain micro-bats are very large too, almost as big as the biggest fruit bats), these emit ultrasonic squeaks for echo-location purposes when flying at night.

Descriptions of the *ahool* and *olitiau* are similar enough to suggest that they belong to the same suborder—but which one?

In view of their size, it would be reasonable to assume, at least initially, that they must surely be mega-bats, and with the *olitiau* in particular there are certain correlations that seem on first sight to substantiate this.

Paul du Chaillu is best remembered as the (in)famous author of grossly exaggerated, lurid descriptions of the gorilla, as encountered by him during his mid-19th century expeditions to tropical Africa.

Less well known is that he was also the discoverer of this vast continent's largest species of bat—a hideous yet harmless fruit bat with a three-foot wingspan known scientifically as *Hypsignathus monstrosus*. On account of its grotesquely swollen, horse-shaped head and an oddly formed muzzle that ends abruptly like the blunt end of a hammer, it is referred to in popular parlance as the hammer-headed or horse-headed bat, and is native to central and western Africa, where it can be commonly found along the larger rivers.

HAMMER-HEADED BAT

This is particularly true during the dry seasons, when the males align themselves in trees bordering the riverbanks and saturate the night air with an ear-splitting Babel of honking mating calls and loud swishing sounds generated by the frenetic flapping of their 18-inch-long wings. A human traveler unsuspectingly intruding upon such a scene as this might well be forgiven for fearing that he had descended into one of the inner circles of Dante's Inferno!

In truth, this hellish gathering is nothing more fiendish than the communal courtship display of the male hammer-heads—each male hoping to attract the attention of one of the smaller, less repulsive females, which select their mates by flying up and down these rows of rowdy suitors, evaluating the potential of each male performer.

On account of its fearsome appearance, large size, and preference for riverside habitats, *H. monstrosus* has been proposed on more than one occasion as a contender for the identity of the *olitiau*. Yet even under the dramatic circumstances surrounding their Cameroon visitation, it is highly unlikely that Sanderson and Russell could have mistaken a bat with a three-foot wingspan for one with a 12-foot spread, especially as they were both very experienced wildlife observers. Moreover, it just so happens that only a few minutes before their winged assailant made its sinister debut, Sanderson had actually shot a specimen of *H. monstrosus*, so the singularly distinctive physical form of this species was uppermost in his mind. Hence, if the mystery beast that appeared just a few minutes later had genuinely been nothing more than a hammer-headed bat, Sanderson would surely have recognized it as such.

Another problem when attempting to reconcile the *olitiau* with this identity is the *olitiau's* belligerent attitude—for in stark contrast, the hammer-headed bat has a widespread reputation as a harmless fruit-eater with a commensurately tranquil temperament. Having said that, however, I must point out that this ostensibly mild-mannered monster does have a lesser-known, darker side to its nature too, one that is much more in keeping with its gargoylesque visage.

As divulged by American zoologist Dr. Hobart Van Deusen (*Journal of Zoology*, 1968), in startling violation of the fruit bat clan's pledge of frugivory, *H. monstrosus* has been exposed on occasion as a clandestine carnivore—because it has been seen to attack domestic chickens, alighting upon their backs and viciously biting them, and it will also carry off dead birds.

Could the *olitiau* therefore be an oversized, extra-savage relative of the hammer-headed bat? In short, an undiscovered giant member of the mega-bat suborder?

Thought provoking though this identity may be, when documenting the subject of giant bats in his book *Investigating the Unexplained* (1972) Ivan Sanderson disclosed that there is also some arresting evidence in support of the converse explanation—namely, that these gigantic forms are in reality enormous micro-bats. This radical notion, which is particularly convincing when applied to the *ahool*, is substantiated by several independent, significant features drawn from these creatures' morphology and lifestyle.

For example, whereas the flattened man-like or monkey-like face of the *ahool* and *olitiau* contrasts markedly with the long-snouted face of most mega-bats, it corresponds well with that of many micro-bats. The presence of both forms of giant mystery bat near to rivers, and the *ahool's* reputedly piscivorous diet, also endorse a micro-bat identity as many species of micro-bat are active fish-eaters, but there is none among the mega-bats. When the *ahool* has been spied upon the ground, or perched upon a tree branch, its wings have been closed up at its sides; a micro-bat trait again—mega-bats fold their wings around their bodies.

Perhaps the most curious, but also the most telling (taxonomically speaking), of the statements by the Sundanese natives concerning the *ahool* is that when observed upon the ground it had stood erect, with its feet turned backwards. To someone not acquainted with bats, all of this may seem nonsensical in the extreme—after all, whoever heard of bats that can stand upright, and which come equipped with back-to-front feet? Yet, paradoxically, these are the very features that vindicate most emphatically the opinion of Sanderson that this mystery creature not only is a bat (rather than a bird or some other winged animal) but is specifically a giant form of micro-bat.

To begin with: although not widely known, micro-bats are indeed capable of standing virtually erect (even though they *mostly* run around on all fours when on the ground); mega-bats, however, never attempt this feat. And switching from feats to feet: those backward-pointing appendages of the *ahool* are worthy of special note too.

There are a few species of bird that habitually hang downwards from branches in a manner reminiscent of bats. These include Asia's *Loriculus* hanging parrots (also called bat parrots for this reason); and, during their courtship displays, the males of New Guinea's blue bird of paradise *Paradisaea rudolphi* and the Emperor of Germany's bird of paradise *P. guilelmi*. If these birds are observed from the front when hanging upside-down, their feet can be seen to wrap around the front portion of the branch, i.e. their toes (excluding the hind ones) curl *away* from the observer, toward the back of the branch.

With bats, however, the reverse is true—their feet wrap around the rear portion of the branch, with their toes curling *towards* the observer. This is because bats' feet really do point backwards—and so, in the event of a bat standing on the ground, its feet would indeed be projecting backwards, precisely as the *ahool's* Sundanese observers have described.

Combining the *ahool's* rearward-oriented feet with its ability to stand upright when on the ground, its identification as an immense micro-bat is thereby lent credence by noteworthy anatomical correspondences—correspondences, moreover, that its native eyewitnesses are unlikely to have appreciated (in terms of their taxonomic implications), and which they would therefore not have incorporated purposefully into their *ahool* accounts if they had merely invented the beast as a hoax with which to fool gullible Westerners.

The possible existence of any type of giant bat with an 11-to-12-foot wingspan, let alone one that is most probably a micro-bat rather than a mega-bat, is bound to raise many an eyebrow and inspire many a derisory sniff within the zoological community. The prospect, furthermore, that there are other such species, in Africa, Madagascar, and elsewhere, too, all successfully evading official discovery and description, may well stimulate even more profound manifestations of incredulity and disdain.

Perhaps, then, these skeptics should pay a visit to western Java and listen for the eerie cry that the wind and good fortune wafted in the direction of Bartels's startled ears; or to western Africa and stand by a silent river to see whether Sanderson's night-winged visitor will swoop out of the evening shadows and into the zoological history books. All of this has happened before — why should it not happen again?

WHEN HOME IS A PILE OF ELEPHANT DUNG

Whereas the previous mystery bats were distinguished by their huge size, the next one is of notably diminutive dimensions—indeed, this is the very characteristic that enables it to indulge in the bizarre day-roosting activity that has incited such scientific curiosity.

It all began in 1955, when John G. Williams, a renowned expert on African avifauna, was taking part in the MacChesney Expedition to Kenya, from Cornell University's Laboratory of Ornithology. In June of that year he encountered Terence Adamson, brother of the late George Adamson of *Born Free* fame, and during a conversation concerning the wildlife inhabiting the little-explored forests of Mount Kulal, in northern Kenya, Adamson casually mentioned a peculiar little bat that had attracted his interest—by virtue of its unique predilection for spending its days snugly concealed inside dry piles of elephant dung, of all things.

Bats are well known for selecting unusual hideaways during the daylight hours, requisitioning everything from birds' nests to aardvark burrows, but there was no species known to science that habitually secreted itself within the crevices present in deposits of elephant excrement. As a consequence, Adamson had been eager to discover all that he could regarding this extraordinary creature.

He had first encountered one of its kind during a walk through Kenya's Marsabit Forest (of which he was warden). After idly kicking a pile of elephant dung lying on the path he was strolling along, he saw a small grey creature fly out of it and alight upon a tree nearby. Expecting it to be nothing more notable than some form of large moth, Adamson was very startled to find that it was an exceedingly small bat, with silver brownish-grey fur, paler upon its underparts. He was especially surprised by its tiny size— its wingspan was even less than that of the familiar pipistrelles, which are among the smallest of bats. Unfortunately, he was only able to observe it for a few moments before it took to the air again and disappeared, but his interest was sufficiently stirred for him to make a determined effort thereafter to seek out other specimens of this odd little animal.

And, as he informed John Williams, during his visit to Mount Kulal he had succeeded in spotting a second one—unceremoniously ejected from its diurnal seclusion when Adamson had kicked over a pile of pachyderm droppings at the base of the Kulal foothills. Unlike the first specimen, however, this one had flown away without making any attempt to land close by, so Adamson had been unable to make any additional observations.

As Williams noted in what seems to be the only account published until now regarding this coprophilic chiropteran (*Animals*, June 1967), he too became very keen to espy, and possibly even capture, one of these elusive denizens of the dung piles in the hope of identifying their species. And so, to his traveling companions' great amusement, Williams made a special point from then on of zealously felling as many dry mounds of elephant excrement as he could on the off chance that he might conjure forth one of these perplexing little bats.

Despite such valiant efforts, however, the elephant dung bat has still not been captured, and its identity remains unresolved—but as John Williams kindly informed me during some correspondence, one species already known to science may provide the answer.

The species in question is a rare micro-bat called *Eptesicus (Rhinopterus) floweri*, first described in 1901 by de Winton, and currently recorded only from Mali and southern Sudan. It is commonly termed the horn-skinned bat, calling to attention the tiny horny excrescences that it bears upon the upper surface of its limbs and tail. This species resembles the elephant dung bat in general size and color, but an important additional reason why Williams favors its candidature as the latter creature's identity is its remarkable preference for day-roosting within holes in the ground, especially among the roots of acacia trees.

As he pointed out to me, this habitat is really quite similar to the crevices and cracks present within dry heaps of elephant dung, hence it is not difficult to believe that this species would utilize these useful sources of daytime roosting sites if such were available. And as the Mount Kulal region of northern Kenya is not only little-explored but also not too far beyond its known distribution range, this provides further reason for looking favorably upon the horn-skinned bat as a realistic answer to the mystery of the elephant dung bat.

THE BLOOD-DRINKING "DEATH BIRD" OF ETHIOPIA

In total contrast to the engaging, if somewhat offbeat, history of the elephant dung bat, the case of the Ethiopian "death bird" is unremittingly macabre and horrific, more akin to the gothic outpourings of Poe and Le Fanu than to anything from the dispassionate, sober chronicles of zoology. Yet in spite of this, it is only too real; at the present time, moreover, it is also unsolved. I am most grateful to Queensland zoologist Malcolm Smith for bringing this chilling but hitherto unexamined case to my attention, and for kindly supplying me with a copy of the original source of information concerning it.

It was during an archaeological expedition to Ethiopia, short-ly before the country was invaded by Italian troops prior to World War II, that Byron de Prorok first learned of Devil's Cave, whose grisly secret he subsequently documented in his travelogue *Dead Men Do Tell Tales* (1943). Journeying through the province of Walaga, he resided for a time at the home of its governor, Dajjazmac Mariam, and while there he was approached by one of the servants, a young boy who began to tell him about a secret cave situated roughly an hour's horseback-ride away, near a place called Lekempti. It was known to the local people as Devil's Cave, and was widely held to be an abode of evil and horror—plagued by devil-men who prowled its darkened recesses in the guise of ferocious hyenas, and by flocks of a greatly feared form of bat referred to as the death bird.

No one had ever dared to penetrate this mysterious cavern, but de Prorok decided to defy its forbidding reputation, because he thought it possible that there would be prehistoric rock paintings inside (especially as its notoriety would have served well in ward-ing off potential trespassers, who might have desecrated any art-work preserved within its stygian gloom).

When de Prorok told his young informant of his decision to visit Devil's Cave, the boy was terrified, but after being bribed with a plentiful supply of gifts he agreed very reluctantly to act as de Prorok's guide—though only on the strict understanding that he would not be held responsible for anything that happened.

The cave was situated high among rocky pinnacles and jungle foliage, but de Prorok succeeded in scrambling up to it, and in removing the several heavy boulders blocking its entrance. Armed with a gun, and leaving his guide trembling with fear outside, he cautiously stepped inside—and was almost bowled over a few min-utes later by a panic-stricken pack of hyenas hurtling down one of the passages to the newly unsealed entrance. Seeking to defend him-self against a possible attack by them, he shot one that approached a little too close, and the echoes from the blast reverberated far and wide, ultimately reaching the ears of two goatherds who came to the cave mouth to find out what was happening. Here they were met by de Prorok, who had followed the hyenas at a respectful distance during their shambolic exit, and who was greatly shocked by the goatherds' pitiful state. They seemed little more than animated skeletons, upon which were hung a few tattered rags.

When, with the boy as interpreter, they learned that de Prorok planned to go back inside the cave, they implored him to change his mind, warning him of the death birds. De Prorok, however, was not afraid of bats and made his way once more through the cave's somber corridors, until he suddenly heard a loud whirring sound overhead. This proved to be a huge cloud of bats, which flew rapidly toward the cave mouth when he fired off a shot in alarm. These, he presumed, must be the dreaded death birds, a line of speculation speedily confirmed when only moments later a rain of bat excrement, dislodged by the shot, began to pelt down upon him from the cave roof, accompanied by an asphyxiating stench that drove him back almost at once to the entrance in search of breathable air.

Outside, he inquired why everyone was so afraid of these bats, to which the two goatherds and the boy all replied that they were blood-suckers, that night after night they came to drink the blood of anyone living near the cave until eventually their unfortunate victims died. This was why the only people living here now were the goatherds (who were forced to do so by the goats' owners). It also explained the goatherds' emaciated state. The death birds' vampiresque activities ensured that none of the goatherds lived very long, but they were always replaced by others, thereby providing the goats with constant supervision—and the death birds with a constant supply of their ghoulish nutriment.

To provide him with additional proof of their statements, the two goatherds took de Prorok to their camp nearby; all of the herders there were equally skeletal—and one was close to death. Little more than a pile of bones scarcely held together by a shroud of ashen skin, this living corpse of a man lay huddled in a cot, with blood-stained

rags and clothes on either side, and was so weakened by the nightly depredations of the visiting death birds that he was unable to
stand, capable only of extending a wraith-like arm. The goatherds
told de Prorok that the death birds settled upon their bodies while
they were asleep, so softly that they did not even wake; and that they
were sizeable beasts, with wingspans of 12 to 18 inches.

As for physical evidence of the death birds' sanguinivorous
nature, the goatherds showed him their arms, which clearly bore a
number of small wounds—the puncture marks left behind by
these winged leeches once they had gorged themselves upon their
hapless hosts?

Nothing more has emerged concerning this gruesome affair, but
for zoologists it would have some significant repercussions if de
Prorok's account could be shown to be perfectly accurate. Only three
species of blood-drinking bat are currently known to science—and all
three of these are confined exclusively to North and South America!

These are the notorious vampire bats, of which the best
known is the common vampire *Desmodus rotundus*, whose range
extends from northern Mexico to central Chile, northern Argentina,
Uruguay, and Trinidad; its numbers have dramatically increased
since the introduction of sheep and other livestock to these areas
with the coming of the Europeans, serving to expand the diversity
and numbers of potential prey victims for it. The other two species
are the white-winged vampire *Diaemus youngi*, recorded from
northeastern Mexico to eastern Peru, northern Argentina, Brazil, and
Trinidad; and the hairy-legged vampire *Diphylla ecaudata*, ranging
from southern Texas to eastern Peru and southern Brazil.

(As a thought-provoking digression, there may also be a
fourth, giant vampire bat in existence. Within the *Proceedings of the
Biological Society of Washington* for December 7, 1988, researchers Drs.
Gary S. Morgan, Omar J. Linares, and Clayton E. Ray formally
described a new species of vampire, 25 percent larger in size than
Desmodus rotundus, based upon two incomplete skulls and skeletal
remains found in Venezuela's famous Cueva del Guácharo, home

of the extraordinary radar-emitting oilbird *Steatornis caripensis*. Dubbed *D. draculae*, this giant vampire bat's remains date from the Pleistocene. However, Brazilian zoologists Drs. E. Trajano and M. de Vivo, in a *Mammalia* paper from 1991, noted that there are reports of local inhabitants in southeastern Brazil's Ribeira Valley referring to attacks upon cattle and horses by large bats that could suggest the continuing survival here of *D. draculae*, although despite extensive recent searches of caves in this area none has been found...so far?)

Over the years, a great deal of misinformation has been dissipated concerning this nocturnal, terror-inducing trio of microbats—including the persistent fallacy that they actively suck blood *out* of wounds, and the equally tenacious yet fanciful misconception that they are enormous beasts with gigantic wings into which they are only too eager to enfold their stricken victims while draining them of their precious blood. In contrast, the truth is (as always) far less exotic and extravagant.

Any creature that can subsist entirely upon a diet of blood (sanginivory) must obviously be highly specialized, and the vampire bats are no exception; Canadian biologist Dr. Brock Fenton from Young University in Ontario has lately suggested that they evolved from bats that originally consumed blood-sucking insects attracted to wounds on large animals, but which eventually acquired a taste for the animals' blood themselves. Yet in overall external appearance these nefarious species are disappointingly mundane—with an unimpressive total length of only two inches to three and a half inches, a very modest wingspan of five to six inches, and a covering of unmemorably brown, short fur. Only when they open their mouth to reveal a distinctive pair of shear-like upper incisors do they display the first intimation of their sinister lifestyle.

These incisors terminate in a central point and have long, scalpel-sharp edges, perfectly adapted for surreptitiously shaving a thin sliver of skin from the body or neck of an unsuspecting (usually sleeping) victim—detected by the vampire's ultrasonic echolocation faculties. The wound that is produced is sufficiently deep to slice through the skin's capillaries, but not deep enough to disturb the victim and thereby waken it (or arouse its attention if already awake); stealth is the byword of the vampire's lifestyle. Aiding the furtive creation of this finely engineered wound are the bat's canine teeth, shorter than the incisors but just as sharp.

Once the wound begins to seep blood in a steady flow, the vampire, delicately clinging to the flank or back of its victim with its wings and hook-like thumbs (*not* with its sharp claws—yet another fallacy), avidly laps the escaping fluid with its grooved, muscular tongue. It can also suck it up by folding its tongue over a notch in its lower lip to yield a tube, but it only sucks blood that has already flowed out of the wound. In addition, its saliva contains anticoagulants, preventing the blood from clotting, and thereby providing the bat with an ample supply (but causing its victim to lose more than would have been the case if the wound had been inflicted by some other type of sharp cutting implement).

Indeed, one of these anticoagulants, plasminogen activator (Bat-PA for short), shows promise as a powerful drug in the prevention of the severe physiological damage caused by heart attacks in humans, according to a study of its effects by research fellow Dr. Stephen Gardell at Merck Sharp & Dohme Research Laboratories in West Point, Pennsylvania.

The vampire's teeth, tongue, and thumbs are not the only specialized facets of its anatomy. Its gut also exhibits some important modifications. Enabling the bat to gorge itself thoroughly before bidding its victim a silent adieu, its stomach has an enormous extra compartment—a tubular, blind-ending diverticulum unattached to the rest of the digestive tract and capable of prodigious distension, rendering it able to hold a voluminous quantity of blood. Sometimes the bat can scarcely fly after feeding, because it is so heavy with freshly ingested blood. Also, its esophagus is specialized for efficient water absorption, a necessity for any obligate sanguinivore because blood contains an appreciable proportion of water.

What all of this means in relation to the Ethiopian death bird is that any bat thriving solely or even predominantly upon a diet of blood is inevitably a much-modified species, rigorously adapted for such a lifestyle—rather than a mere opportunist species that in certain localities has switched (through some unusual set of circumstances) from its normal diet to a sanguinivorous existence. In other words, if de Prorok's account is a truthful one, then surely the death bird must be a species new to science? After all, there is currently no *known* species of Old World bat that is a confirmed blood-drinker. This, then, is plainly one plausible answer to the death bird mystery—but it is not the only such answer.

I am exceedingly grateful to the late John Edwards Hill, bat specialist and formerly Principal Scientific Officer at the British Museum (Natural History), who presented me with a great deal of information that offers a completely different outlook upon this perplexing case. It is well known that the New World vampire bats transmit livestock diseases from one animal victim to another, in a manner paralleling the activities of mosquitoes and other sanguinivorous insect vectors. They also carry rabies to man, though this is a much rarer occurrence than the more lurid reports in the popular press would have us believe. Moreover, bats of many species all around the world are known to contract many different types of bacterial, viral, and protozoan diseases, which can be spread to other organisms via parasites such as body lice and ticks that live upon the bats' skin or fur. Relapsing fever in man, for example, is caused by the bacterium *Borrelia recurrentis,* carried by lice and ticks that have in turn derived it from former rodent or bat hosts.

Accordingly, during our communications concerning the death bird, Hill suggested that it is possible that humans venturing in or near a cave heavily infested with bats (like Devil's Cave, for instance) would become infected with such diseases—if lice or ticks, dropping from the bats as they flew overhead, bit the unfortunate humans upon which they landed. A parasite-borne infection of this nature would account for the bite-like wounds of the goatherds observed by de Prorok; and, depending upon the precise type of infection, could ultimately give rise to the emaciated condition exhibited by these afflicted persons.

Additionally, native superstition and a deep-rooted fear of bats might be sufficient, when coupled with the distressing effects of a parasite-borne infection, to nurture the belief among such poorly educated people as these that they were the victims of blood-sucking bats—the notion of vampirism is very ancient and widespread in human cultures worldwide (the Maya of pre-Columbian Mesoamerica even worshipped the vampire bat as a god named Camazotz).

Two other medical explanations for the death bird case were also raised by Hill during our correspondence (though he rated both of these as being less plausible than the likelihood of a parasite-borne disease's involvement).

As Devil's Cave contained large quantities of bat excrement, perhaps these droppings harbored the spores of the soil fungus *Histoplasma capsulatum* (even though this is more usually associated with bird guano); if inhaled, these spores can cause an infection of the lungs known as histoplasmosis, which *can* prove fatal (but severe cases are not common).

Alternatively, an illness called Weil's disease also offers some notable parallels with the "death bird syndrome." Also referred to as epidemic spirochaetal jaundice and as leptospirosis ictero-haemorrhagica, Weil's disease is caused by spirochaete bacteria of the genus *Leptospira*, and is usually spread by rodents, but the bacteria have been found in a few species of bat as well. Infection generally occurs through infected drinking water, and among the ensuing symptoms of contraction is the appearance of small hemorrhages in the skin, which could be mistaken for bites. Also, the accompanying damage to the kidneys and liver, jaundice, and overall malaise experienced by sufferers could explain the goatherds' haggard, wasted form.

Clearly, then, the case of the dreaded death bird and the stricken herders is far from being as straightforward as it seemed on first sight, and may involve any one, or perhaps even more than one, of the above solutions. An aspect of the case that is self-evident, however, is the necessity for a specimen of the death bird to be collected and formally studied. Only then might the resolution of this mystifying and macabre cryptozoological riddle be finally achieved—but in view of the perennially uncertain political climate associated with Ethiopia in modern times, even this is unlikely to prove an easy task to accomplish.

Until then, the secret of this purportedly deadly unidentified creature will remain as dark and impenetrable as the grim cave from which its winged minions allegedly issue forth each night to perform their vile abominations upon the latest tragic campful of doomed, defenseless goatherds.

THE *SASABONSAM* SCULPTURE

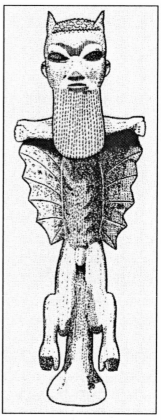

One particularly bizarre entity deserves brief mention here if only because it is as likely to have been inspired by encounters with strange bats as by anything else. In an absorbing *West African Review* article for September 1939, J.B. Danquah included a photograph of a carving produced by a member of Ghana's Asantehene people, which depicted a grotesque being—referred to as the *sasabonsam*. According to the carving, the *sasabonsam* has a human face with a broad, lengthy beard, long hair, and two horns or pointed ears; an extremely thin body with readily visible ribs; short stubby forelimbs that,

SASABONSAM SKETCH BASED ON DANQUAH'S PHOTOGRAPH

when lifted upwards, reveal a pair of thin membranes resembling a bat's wings; and long, twisted legs terminating in feet with distinct toes. Most authorities dismiss this creature as a myth, but a few have wondered whether it could be some form of giant bat.

Of particular interest to Danquah was the narrative of a youth present in a crowd of people (which included Danquah himself) observing this carving being photographed, because the youth averred that once, at his own Ashanti town, he had seen the body of a *sasabonsam*—killed by a man called Agya Wuo, who had brought its carcass into the town. The youth confirmed that the carving was very like the *sasabonsam* that he had seen, and he added some details based upon the dead specimen.

This had been more than five and a half feet tall; its wings were very thin but when fully stretched out could yield a span of up to 20 feet; its arms (unlike the carving's) were very long, as were its teeth; its skin was spotted black and white; and there had been scaly ridges over its eyes, with hard, stiff hair on its head. The man who had killed it had stated that he had encountered it sleeping in a tree hollow, and that it had emitted a cry like that of a bat but deeper. The body was allegedly taken to the bungalow of L.W. Wood, the region's District Commissioner, who photographed it on February 22, 1928.

Danquah later contacted Wood, who seemed uncertain whether he had indeed photographed such a creature; in addition, he claimed that although he was in Ashanti in February 1918, he was not there in February 1928. It is possible, therefore, that if this account is true, the youth who informed Danquah had become confused regarding the date.

In any event, neither the carving nor the description of the dead specimen bear more than the most passing of resemblances to a bat—but it is equally true that they do not call to mind any other type of animal either! If the fate of the dead specimen (always assuming of course that there really was such a specimen) could be ascertained, its skeleton would solve the currently insoluble riddle of the *sasabonsam's* identity—but until then, this "bat with a beard" must remain just another one of the varied beings of African mythology that appear to have no known counterpart in the zoological world.

IS BATMAN ALIVE AND WELL AND LIVING ON THE ISLAND OF SERAM?

Seram (also spelled "Ceram") is the second largest island of the Moluccas group in eastern Indonesia, west of New Guinea, but the dense rainforests occupying its interior are still largely unexplored—which is why the following communication to me from cryptozoological explorer Bill Gibbons is one of the most remarkable that I have ever received. I am very grateful to Bill, formerly of the U.K. but now living in Canada, for sharing this data with me, and permitting me to document it here.

It was gathered by one of Bill's colleagues, tropical agriculturalist Tyson Hughes, who visited Seram in June 1986, working there for the VSO (Voluntary Service Overseas) as the project manager of a model farm with 16 staff members. According to Bill's communication to me:

> During his free time, Tyson became interested in native accounts of a bizarre creature known as the Orang Bati, meaning 'flying men' in Indonesian. The villagers of the coastal regions live in literal terror of the Orang Bati, which, they claim, live in the interior of long-dead volcanic mountains. At night the Orang Bati will leave their mountain lairs and fly across the jungles to the coast, where they seize human victims from the coastal villages and carry them back to the mountains.
>
> Although the local police are aware of the native accounts of the Orang Bati, they dismiss them out of hand, mocking any reports that filter their way

into civilisation. The majority of serving police
officers are actually Javan, so their skepticism is,
perhaps, not unexpected. However, the fear of the
Orang Bati is considerable among almost all the
indigenous inhabitants of Seram Island. Those
abducted by the Orang Bati are never seen again.
These mysterious creatures are described as
human in form, with red coloured skin, black
wings, and long, thin tails. Tyson was able to per-
suade several village hunters to lead him through
a dense area of jungle towards the alleged lair of
the Orang Bati. Although he became the first
white man to explore part of Seram Island, the
Orang Bati remained elusive.

In a later letter to me, Bill added that the *orang bati* emits a long,
mournful wail, is four feet to five feet tall, prefers to abduct infants
and children, and terrorizes the coastal village of Uraur among others. He also noted that there do not appear to be any monkeys on
Seram—which, if true, could explain why it preys upon human
youngsters.

On first sight, if we assume that the *orang bati* is indeed real
and not a figment of superstition-inflated imagination, the most reasonable solution to the mystery of its identity is to nominate the
existence of a scientifically unrecorded species of giant bat—
though any bat large and powerful enough to abduct children
would be both a marvel and a monster.

However, an even more radical, controversial comparison vig-
orously vies for my attention whenever I attempt to reconcile the
orang bati with the identity of a giant bat. According to the testimony
of three U.S. Marines, one of whom was Earl Morrison, while on
guard duty near Da Nang in South Vietnam during summer 1969
they spied a mysterious entity flying slowly towards them that possessed bat-like wings, but it was far larger than any normal bat. As

it came nearer, the trio realized to their amazement that this bizarre being resembled a naked woman about five feet tall, with black furry skin and black wings—to which were attached her arms, hands, and fingers. After "she" had flown overhead, and traveled about 10 feet beyond them, the men could hear her wings flapping.

Inevitably but not unreasonably, this report has been dismissed in the past by skeptics as a misidentification of some large species of bird, or bat, resulting from long-term stress induced by living in a war zone. In view of the *orang bati*, however, the time may have come to re-examine this mystifying account from southeast Asia in a new light.

Few creatures are likely to stay hidden from human interference as effectively as those that are nocturnal, winged, and capable of eliciting a profound degree of irrational fear and horror in mankind. Little wonder, therefore, that bats remain some of the most mysterious mammals alive today, with new species appearing—and even supposedly extinct species reappearing—almost every year.

Consequently, the prospect that certain of the more remote, inhospitable reaches of the world could harbor very distinctive bats wholly unknown to science is really not very surprising—on the contrary, it would only be very surprising if this were *not* the case.

CHAPTER 6

Man-Eating Trees and Vampire Plants

It may well be that there existed or still exists an unknown relatively large carnivorous plant that has one or two of the adaptations described for trapping birds or other smaller arboreal creatures. There are still large forest areas, especially in the southeastern and south-central portions of Madagascar, that would be interesting to explore.

Roy P. Mackal—*Searching For Hidden Animals*

WHEN THE VENUS FLYTRAP *DIONAEA MUSCIPULA* WAS first made known to botanists in the 1760s, they would not believe that it could actually catch and consume insects—until living specimens were observed in action. Moreover, reports have also emerged from several remote regions of the world concerning horrifying carnivorous plants that can ensnare and devour creatures as large as birds, dogs, and monkeys—and sometimes even humans! Once again, however, these accounts have been received with great skepticism by science—relegating them to the realms of fantasy alongside such fictitious flora as Audrey II, the bloodthirsty "Green Mean Mother" star of the cult movie musical *Little Shop of Horrors* (1986). But could such botanical nightmares really exist? Examine the following selection of remarkable reports, and judge for yourself.

THE MAN-EATING TREE OF MADAGASCAR

Perhaps the most incredible case on file is one that first came to Western attention via an extraordinary letter received in 1878 by Polish biologist Dr. Omelius Fredlowski. According to the letter's contents, at least one Western explorer claimed to have witnessed an all-too-real, fatal encounter with a rapacious botanical monster that would put even the worst excesses of Audrey II to shame!

The letter was from Carl Liche, a German explorer who had been visiting a primitive tribe called the Mkodos on the African island of Madagascar. While there, he and a fellow Westerner called Hendrick were shown a grotesque tree, to which humans were sacrificed:

> If you can imagine a pineapple eight feet high and thick in proportion resting upon its base and denuded of leaves, you will have a good idea of the trunk of the tree, a dark dingy brown, and apparently as hard as iron. From the apex of this truncated cone eight leaves hung sheer to the ground. These leaves were about 11 or 12 ft long, tapering to a sharp point that looked like a cow's horn, and with a concave face thickly set with strong thorny hooks. The apex of the cone was a round white concave figure like a smaller plate set within a larger one. This was not a flower but a receptacle, and there exuded into it a clear treacly liquid, honey sweet, and possessed of violent intoxicating and soporific properties. From underneath the rim of the undermost plate a series of long hairy green tendrils stretched out in every direction. These were 7 or 8 ft long. Above these, six white almost transparent palpi [tentacles] reared themselves toward the sky, twirling and twisting with a marvellous incessant motion. Thin as reeds, apparently they were yet 5 or 6 ft tall.

Suddenly, after a shrieking session of prayers to this sinister tree, the natives encircled one of the women in their tribe, and forced her with their spears to climb its trunk, until at last she stood at its summit, surrounded by its tentacle-like palpi dancing like snakes on all sides. The natives told the doomed woman to drink, so she bent down and drank the treacle-like fluid filling the tree's uppermost plate, and became wild with hysterical frenzy:

> But she did not jump down, as she seemed to intend to do. Oh no! The atrocious cannibal tree that had been so inert and dead came to sudden savage life. The slender delicate palpi, with the fury of starved serpents, quivered a moment over her head, then fastened upon her in sudden coils round and round her neck and arms; then while her awful screams and yet more awful laughter rose wildly to be instantly strangled down again into a gurgling moan, the tendrils one after another, like green serpents, with brutal energy and infernal rapidity, rose, retracted themselves, and wrapped her about in fold after fold, ever tightening with cruel swiftness and the savage tenacity of anacondas fastening upon their prey. And now the great leaves slowly rose and stiffly erected themselves in the air, approached one another and closed about the dead and hampered victim with the silent force of a hydraulic press and the ruthless purpose of a thumb screw.
>
> While I could see the bases of these great levers pressing more tightly towards each other, from their interstices there trickled down the stalk of the tree great streams of the viscid honeylike fluid

mingled horribly with the blood and oozing vis-
cera of the victim. At the sight of this the savage
hordes around me, yelling madly, bounded forward, crowded to the tree, clasped it, and with
cups, leaves, hands and tongues each obtained
enough of the liquor to send him mad and frantic.
Then ensued a grotesque and indescribably
hideous orgy. May I never see such a sight again.

The retracted leaves of the great tree kept their
upright position during ten days, then when I
came one morning they were prone again, the
tendrils stretched, the palpi floating, and nothing
but a white skull at the foot of the tree to remind
me of the sacrifice that had taken place there.

Liche was not the only visitor to Madagascar to learn of this
nightmarish species. Chase Salmon Osborn, governor of Michigan
from 1911 through 1913, journeyed to Madagascar during the
early 1920s in the hope of seeing for himself the terrible tree. Sadly
for science (but perhaps fortunately for him!), he did not succeed in
locating one, but he discovered that it was well known to natives all
over the island, and even some of the Western missionaries working there believed in its existence. He also learned that from the very
earliest times Madagascar had been known as "the land of the man-
eating tree."

Nevertheless, there is much to doubt in Liche's testimony
regarding this herbaceous horror—not least of which is whether
Liche himself ever existed! Eminent biochemist and cryptozoologist
Dr. Roy P. Mackal, now retired from the University of Chicago,
devoted an entire chapter to the Madagascan man-eating tree in his
book *Searching For Hidden Animals* (1980), but was unable to discover
any background history concerning Liche, and even the original
publication source of Liche's letter remains a mystery.

No less controversial is the morphology of the man-eating tree, for Liche's description brings together an extraordinary (and highly unlikely) collection of specialized structural features seemingly drawn from several wholly different, unrelated groups of plants. As Mackal justifiably pointed out, such an amazing combination of characteristics could not reasonably be the outcome of effective evolutionary adaptation. Moreover, its ever-animate, writhing palpi are unlike any structure ever reported from *any* known species of plant.

Consequently, Mackal dismissed the existence of Madagascar's man-eating tree, at least in the form attributed to it by Liche. However, as he noted when concluding his chapter dealing with this lethal entity, Liche's description may be a highly embellished, exaggerated account of a real but smaller, less dramatic species of carnivorous plant native to Madagascar. Certainly, any region of the world that has offered up for scientific scrutiny as many truly unique, endemic species as this veritable island continent has must surely retain the potential for concealing some major biological surprises even today.

Indeed, there may even be some photographic evidence for the existence of such a plant. Czech explorer Ivan Mackerle is probably best known in cryptozoological circles as the most famous modern-day seeker of the Mongolian death worm (see Chapter 2). However, he also has a longstanding interest in stories of mysterious flesh-eating flora, and in 1998 he led a month-long expedition to Madagascar in order to investigate reports of the man-eating tree, which he refers to as the *tepe*. Moreover, in a letter to me concerning this, Ivan included a truly tantalizing snippet of information of which I was not previously aware.

In 1935, a former British army officer called L. Hearst apparently spent four months in Madagascar, and while there he took photographs of some unknown species of tree under which lay the skeletons of various sizeable animals. According to Ivan, these photos were later published somewhere, but he has been unable to find out where. Some scientists who saw the photos claimed that they were fakes, so Hearst returned to Madagascar to obtain more convincing proof, but died in mysterious circumstances.

This, at least, is the story that Ivan has pieced together, but he has been unable as yet to provide conclusive corroboration for it. So if anyone reading this book has any relevant information, or knows where the tree photos were published, I'd be very interested to receive details. As for Ivan's own Madagascar expedition, he was unable to uncover any evidence in support of Liche's claims. In a letter to me of September 21, 1998, Ivan wrote:

> We had taken a Malagasy guide and interpreter with us, who lives in Prague and knows Czech. And so we could speak with the natives about mysteries. We had travelled all over the country, mainly in the south region. It is interesting, but no-one had known anything about the man-eating tree. Neither people in town (botanists, journalists, etc) nor natives. They had heard only about pitcher plants. Natives know killer trees but no man-eating ones. The story of Karl [sic] Liche is unknown there. We spoke with many botanists. I could not believe it, because I had supposed that it was a widespread legend there. But killer trees are also very interesting. Many of them are little-known or unknown to science. We found the killer tree 'kumanga', which is poisonous when it has flowers. We took gas-masks for protecting ourselves, but the tree did not blossom at that time. We had seen a skeleton of a dead bird and a dead turtle [tortoise] under the tree. The tree grows only in one place in Madagascar and it is rare today. It was difficult to find it.

So it would appear that even though the man-eating tree is non-existent, Madagascar can still tantalize mainstream botany, courtesy of the *kumanga* killer tree. As Ivan's team encountered it, this mystifying species clearly exists—but what can it be, and is it truly capable of achieving the lethal effects claimed by the local people? Mindful that a number of harmless plants on Madagascar have been accredited with all manner of sinister talents in Malagasy folklore, it would hardly be surprising or unprecedented if the *kumanga's* deadly tendencies owe more to imaginative fiction than biological fact. Conversely, I have so far been unable to determine the *kumanga's* taxonomic identity—could it therefore be unknown to science, echoing Ivan's above-quoted words? There are evidently some notable mysteries of the cryptobotanical kind still awaiting resolution on the exotic island of Madagascar.

Moreover, in a highly unexpected twist to the long-running saga of Madagascar's putative man-eating tree, Canadian researcher W. Ritchie Benedict revealed in 1995 that he had uncovered a published but hitherto-unpublicized newspaper account (*The Watchman*, New Brunswick, May 29, 1995) regarding this cryptobotanical wonder dating back to 1875 (hence predating the Liche letter by three years), and which indicates an origin for it not in Madagascar but in New Guinea! How ironic it would be if the reason why the greatest mystery plant of all time has never been scientifically exposed is that everyone has been looking for it on the wrong island!

CENTRAL AMERICA'S *YA-TE-VEO* TREE AND BRAZIL'S DEVIL TREE

A similar carnivorous plant to Madagascar's supposed man-eating tree, and also currently undiscovered, is said to exist in parts of Central America. In *Sea and Land* (1887), J.W. Buel called it the *ya-te-veo* tree, and described it as having a short, thick trunk with immense spine-like shoots at its summit that bear dagger-like thorns along their edges. These shoots hang down to the ground, and appear lifeless—until an unwary person walks between them, toward the trunk itself. Then, without warning, the shoots rise up and entwine themselves around him, pressing him onto the trunk's surface, upon which they instantly impale him with their long thorns and crush him until his body is drained with blood, which is rapidly absorbed through the tree's surface. Once again, however, the inordinate motility of this plant's shoots poses a major dilemma when attempting to assess its credibility.

Equally problematic is the Brazilian devil tree, alluded to by Harold T. Wilkins in *Secret Cities of Old South America* (1952) and also termed the octopod tree. Native to the Mato Grosso, it is said to be as big as a willow, but hides its branches deep in the earth or the surrounding undergrowth. Should anything (or anyone) go near to it or trip over its concealed branches, however, this diabolical plant will stealthily draw them out from under the soil or bushes and snare its unwary victim in the grip of their ever-tightening tendrils.

During the summer of 1932, Captain Thomas W.H. Sarll from Middlesex in England set off for the Amazon on a courageous quest to locate, uproot, and return home to England with a living octopod tree. As nothing was heard of his expedition's outcome, however, we can only assume that it did not accomplish its bold objective.

NICARAGUAN DOG-DEVOURING TREE

The following item appeared in the column "Science Jottings" by naturalist Dr. Andrew Wilson for the *Illustrated London News* on August 27, 1892:

> I have lately met with the description of a very singular plant, given originally, I believe, in a provincial newspaper…It appears that a naturalist, a Mr. Dunstan by name, was botanising in one of the swamps surrounding the Nicaragua Lake. The account goes on to relate that "while hunting for specimens he heard his dog cry out, as if in agony, from a distance. Running to the spot

whence the animal's cries came, Mr. Dunstan
found him enveloped in a perfect network of
what seemed to be a fine, rope-like tissue of roots
and fibres. The plant or vine seemed composed
entirely of bare, interlacing stems, resembling
more than anything else the branches of a weep-
ing willow denuded of its foliage, but of a dark,
nearly black hue, and covered with a thick, viscid
gum that exuded from the pores. Drawing his
knife, Mr. Dunstan attempted to cut the poor
beast free, but it was with the very greatest difficulty that he managed to sever the fleshy muscular fibres [sic] of the plant. When the dog was
extricated from the coils of the plant, Mr. Dunstan
saw to his horror that its body was bloodstained,
while the skin appeared to be actually sucked or
puckered in spots, and the animal staggered as if
from exhaustion. In cutting the vine the twigs
curled like living, sinuous fingers about Mr.
Dunstan's hand, and it required no slight force to
free the member from their clinging grasp, which
left the flesh red and blistered. The tree, it seems,
is well known to the natives, who relate many
stories of its death-dealing powers. Its appetite is
voracious and insatiable, and in five minutes it
will suck the nourishment from a large lump of
meat, rejecting the carcass [sic] as a spider does
that of a used-up fly." This is a very circumstan-
tial account of the incident, but in such tales it is,
of course, absurd "to leave such a matter to a
doubt." If correct, it is very clear we have yet to
add a very notable example to the list of plants
which demand an animal dietary as a condition
of their existence; and our sundews, Venus fly-

traps, and pitcher plants will then have to "pale their ineffectual fires" before the big devourer of the Nicaragua swamps.

MEXICAN SNAKE-TREE

What makes the Nicaraguan "dog devourer" of particular note is that reports of a similar entity have also emerged further north, where it is referred to as the Mexican snake-tree. Once again, it featured in Wilson's column, within an item for September 24, 1892:

> The "snake-tree" is described in a newspaper paragraph as found on an outlying spur of the Sierra Madre, in Mexico. It has movable branches (by which, I suppose, is meant sensitive branches), of a "slimy, snaky appearance," which seized a bird that incautiously alighted on them, the bird being drawn down till the traveller lost sight of it. Where did the bird go to? Latterly it fell to the ground, flattened out, the earth being covered with bones and feathers, the débris, no doubt, of former captures. The adventurous traveller touched one of the branches of the tree. It closed upon his hand with such force as to tear the skin when he wrenched it away. He then fed the tree with chickens, and the tree absorbed their blood by means of the suckers (like those of the octopus) with which its branches were covered.

> I confess this is a very "tall" story indeed. A tree which is not only highly sensitive, but has branches which suck the blood out of the birds it captures, is an anomaly in botany. Mexico is surely not quite such an unexplored territory that

a tale like this should go unverified. I give the
story simply for what it is worth. A lady correspondent, however, reminds me that in Charles
Kingsley's West Indian papers he describes a similar tree to the Nicaraguan plant of Mr. Dunstan.

A number of points can be made in relation to Wilson's
above comments. First, it seems strange that whereas he is highly
skeptical of the snake-tree's alleged capabilities, he seems to have
rather less difficulty in accepting these selfsame talents for the
Nicaraguan plant. Indeed, judging from the similarities in the
descriptions of these two plants, it seems reasonable to assume that
they represent closely related species. In reality, Wilson's doubt concerning the Mexican plant seems to stem not so much from its func-
tional morphology as from its geographical locality. Yet as
cryptozoologists will readily point out, even today Mexico can present some notable mysteries for the naturalist. This is best exemplified by the onza. This elusive cheetah-like cat's existence has been
denied by zoologists for more than three centuries, but it has frequently been reported by native Mexicans, and alleged onza skulls
and specimens have also been collected. Moreover, the newspaper
account referred to by Wilson is not the only source of data regarding a vampiresque Mexican mystery plant.

Despite undertaking a detailed bibliographical search, I have
so far been unable to locate that particular newspaper article.
However, during my search I did uncover a more recent account,
which seems to refer to a comparable species.

In early 1933, the French explorer Baron (sometimes designated as Comte) Byron Khun de Prorok led a party of explorers into
the almost impenetrable jungle region of the Chiapas, in southern
Mexico, and in the August 1934 issue of *Wide World Magazine* he
recounted some of the more memorable episodes that occurred during this journey. The first of these is the one of interest to us, which
took place just two hours after first entering the jungle, by which

time its omnipresent humidity had begun to take its toll:

> After two hours' march we breathed with diffi-
> culty, and were bathed in perspiration.

> Suddenly I saw Domingo, the leader of the
> guides, standing before an enormous plant and
> making gestures for me to go to him. I wondered
> what could be the matter.

> I soon saw; the plant had just captured a bird!
> The poor creature had alighted on one of the
> leaves, which had promptly closed, its thorns
> penetrating the body of the little victim, which
> endeavoured vainly to escape, screaming mean-
> while in agony and terror.

> *"Plante vampire!"* explained Domingo, a cruel
> smile spreading over his face.

> Involuntarily I shuddered; the forest was casting
> its evil spell upon me.

It is quite possible that this "vampire plant" and the snake-tree (and perhaps even the Nicaraguan example too?) are the same species, their differences in morphology due not to nature but to inaccurate reporting. De Prorok was a renowned explorer, hence it is probably best to look upon his description as being more accurate than the earlier newspaper account. Certainly, it would be easier to accept the existence of a plant with sensitive thorn-bearing leaves functioning somewhat like those of the Venus flytrap than the evolution of botanical devices bearing suckers capable of actively draining a bird of its blood.

Unlike animals, plants do not possess a nervous system.

However, they do generate electrical signals that greatly resemble nerve impulses. Recordings of electrical signals (action potentials) generated by the Venus flytrap were first obtained as long ago as 1873, by London physiologist Sir John Burdon-Sanderson—when he bent one of the trigger-hairs on one of this carnivorous plant's leaves, he recorded the passage across that leaf of an electrical signal traveling at 20 mm per second. It has since been shown that if two such signals are generated (i.e. by something brushing against one or more of the leaf's trigger hairs) within 20 seconds, these stimulate the leaf's closure (it folds along its midrib, like a book closing), effecting the flytrap. This is an extremely rapid movement (for a plant), taking only a second or so; the movement itself is engineered via extremely rapid growth of the leaf, triggered by a fall in the pH of the leaf's cell walls (i.e. by an increase in their acidity).

In the famous "sensitive plant" *Mimosa pudica,* electrical impulses are propagated throughout any leaf that is touched or brushed against. These impulses cause each of the leaflets comprising that leaf to fold up—another notable rapid movement taking no more than a second, but this time created by turgidity changes in specialized regions of the leaf called pulvini (absent in Venus flytraps).

A mechanism involving the touch-induced generation of electrical impulses stimulating the rapid closure of thorn-bearing leaves is indicated by de Prorok's description of the Mexican "vampire tree," and is certainly well within the capability of plant evolution.

These deadly entities may also be akin to a second carnivorous mystery plant from Brazil that was noted by Harold T. Wilkins in *Secret Cities of Old South America* (1952). Like his earlier-mentioned octopod (devil) tree, Wilkins's second species is supposedly equipped with motile tentacular branches. Unlike the octopod tree, however, this species' whereabouts can be readily detected by humans because it releases a smell resembling the stink of a rotting carcass. Its preferred prey consists of small birds, which it lures with sweet-tasting berries. When an unsuspecting bird alights upon it to peck at its berries, one of the tree's tentacles seizes one of its wings immediately, and unless the bird can tear itself away it is pulled up against the tree's trunk, held in place by suckers, and crushed to death—after which its blood is absorbed by the suckers before its dry, lifeless carcass is dropped to the ground.

BRAZILIAN MONKEY-TRAP TREE

Also reported from Brazil, and even more formidable, is the monkey-trap tree. In his book *Carnivorous Plants* (1974), Randall Schwartz refers to a "recent report" (but without providing any reference for this) credited to a Brazilian explorer called Mariano da Silva. According to Schwartz's rendition of this report, da Silva had been searching for the settlement of Yatapu Indians on the Brazilian border with Guyana when he encountered an animal-devouring tree. It releases a very distinctive scent that is particularly attractive to monkeys, luring them to it and enticing them to climb its trunk—whereupon its leaves totally envelop them, rendering these hapless creatures invisible and inaudible to anyone witnessing this grisly spectacle. Three days or so later, the leaves open again, and from them drop the bones of their victims, from which every vestige of flesh has been stripped.

If genuine, the mechanism involved once again compares to that of the Venus flytrap, but the leaves must be much tougher and extraordinarily inflexible to be capable of suppressing the struggles of creatures as large as monkeys.

INDIAN MOUSE-EATING PLANT

Rather more plausible than a monkey-ingesting tree is a plant capable of consuming mice. After all, as noted by Hal Bowser, some pitcher plants are so large that they can not only trap insects and spiders but even tiny mammals and small snakes, too. However, there may well be some mouse eaters still eluding official discovery and description—as in the case of the cryptic rodent remover supposedly exhibited for a while in London. As an example of other carnivorous plants to substantiate his detailed account of Madagascar's man-eating tree, Chase Salmon Osborn included the following account in his book *Madagascar, Land of the Man-Eating Tree* (1924):

> At the London Horticultural Hall in England there is a plant that eats large insects and mice. Its principal prey are the latter. The mouse is attracted to it by a pungent odor that emanates from the blossom which encloses a perfect hole just big enough for the mouse to crawl into. After the mouse is in the trap bristle-like antennae infold it. Its struggles appear to render the Gorgonish things more active. Soon the mouse is dead. Then digestive fluids much like those of animal stomachs exude and the mouse is macerated, liquefied and appropriated. This extraordinary carnivorous plant is a native of tropical India. It has not been classified as belonging to any known botanical species.

Of all the mystery carnivorous plants documented here, this seems to be the most feasible one. Its mode of operation recalls that of the familiar pitcher plants. These species possess a tall vertical pitcher-shaped receptacle into which tiny animals are lured by sweet nectar secretions on the outside of the pitcher, but once inside they are unable to crawl out again, due to the pitcher's slippery internal wall and inward-curving spines at its brim, and so they fall down to the bottom, into a pool of fluid containing digestive enzymes, where they swiftly drown and become digested. In the case of the mouse-eating plant, a noteworthy modification has been added to this mechanism, whereby the trap's internal wall bears retaining spines that are reminiscent of the external but functionally comparable tentacles of another group of carnivorous plants, the sundews.

PITCHER PLANT

DEATH FLOWER OF THE SOUTH PACIFIC

An inordinately deceitful plant of prey is a monstrous man-eater that became known to a Captain Arkright, an explorer, in 1581. According to South Pacific folklore, it existed on one of a group of islets encircled by a coral ring in this region of the world. The islet in question was called El Banoor ("Island of Death") on account of its sinister inhabitant. The plant was huge, with enormous, brightly hued petals that opened out to yield a seductive, scented chamber whose soothing colors and bewitching fragrance beguiled

THE BEASTS THAT HIDE FROM MAN

many an unsuspecting human, tempting him to enter and rest upon its lowermost petals, and thence to close his eyes and sleep—a sleep from which he would never awaken. For as he slumbered, the plant's petals would envelop him and bathe his unresisting form in fiery acidic juices, digesting his body even as its narcotic fragrance retained its all-embracing shroud around his mind.

In his book *Myths and Legends of Flowers, Trees, Fruits, and Plants* (1911), Charles M. Skinner dismisses this report as nothing more than a greatly exaggerated early account of a Venus flytrap! However, this unmistakable species is wholly confined in the native state to a single large bog around Wilmington, North Carolina (a bog which, intriguingly, just happens to reside in what is thought by some to be an ancient meteorite crater!).

Conversely, Dr. Roy P. Mackal wondered whether it could have been a distorted account of *Rafflesia arnoldii* from Sumatra, a parasitic species that sends root-like structures resembling the mycelia of fungi into the roots of another plant, a vine called *Tetrastigma*, in order to draw up nourishment. What makes *R. arnoldii* so famous in biological circles is its flower—huge, flat, and at least three feet across, constituting the largest flower in the world. It is composed of several crimson petals, leathery in texture and speckled with cream blotches, plus a central bowl containing spike-like structures and several pints of dirty water.

However, because this flower is horizontal, not vertical, it does not correspond with the South Pacific death flower's structure as described to Arkright. It seems that although *Rafflesia* could be stepped *upon*, it could not be stepped *into*.

Furthermore, there is another major problem when seeking to reconcile this controversial entity with *Rafflesia*. Far from releasing a pleasant, alluring fragrance capable of enticing humans to their doom, *Rafflesia* releases a hideous, stomach-churning odor redolent of rotting flesh—thereby repelling humans, but proving irresistible to blowflies, which crawl upon its unappealing form in a futile search for decaying meat. In so doing, they become covered in pollen, which they subsequently transfer to other specimens of *Rafflesia* when similarly lured, thus effecting pollination of this grotesque giant.

The identity of the death flower (if ever more than a traveler's tale to begin with) thus remains unresolved. And so too, incidentally, does that of an unconfirmed, but equally sinister, slumber-inducing plant reputedly known as *el juy-juy* to the natives inhabiting the Chaco forest on the border of Argentina and Bolivia. In his above-cited book *Secret Cities of Old South America* (1952), Harold T. Wilkins included the following account:

> The plant is one of great beauty and seductiveness and is said to exhale a soporific perfume which sends to sleep men or large animals unlucky enough to seek its shade, in the noonday and siesta hours when the denizens of the forest are silent. Once the victim has sunk into a drugged sleep, the floral canopy overhead sends down masses of lovely blossoms, each flower of which is armed with a powerful sucker, which draws from the body all its blood and juices, leaving not even a fragment to tempt the vulture to shoot down from the skies to gorge on a bare skeleton.

Such a chillingly intelligent plant would make a spellbinding star of some imaginative science fiction film or novel, or even—as is most likely to be its true origin—a traditional jungle legend, but as a plausible plant of real-life botany, it leaves a lot to be desired!

FINAL THOUGHTS

In an article contemplating cryptozoology and its *raison d'être* (*Cryptozoology,* 1987), Russian biologist Dr. Dmitri Bayanov opined that there is no botanical equivalent, i.e. a science of "cryptobotany," nor would there be one in the future, because whereas animals remain hidden by virtue of being actively able to hide themselves, plants cannot do this.

However, this attitude is at variance with the definition of cryptozoological subjects as coined by Dr. Bernard Heuvelmans—who has stated that such animals are ones that are known to local inhabitants (natives or colonists) of their territory, but have so far remained undiscovered by science (despite specific searches in some cases).

As shown by the controversial organisms discussed in this chapter, there may also be some notable plants that fit this description. To my mind, the search for these is surely no less important, and their ultimate disclosure no less exciting, than the seeking and discovery of unknown animals. So perhaps there is scope for a sibling science, that of cryptobotany, after all—with its controversial plants of prey unquestionably among its most intriguing and compelling prizes on offer.

CHAPTER 7

Hairy Reptiles and Furry Fish

In her abnormalities nature reveals her secrets.

JOHANN WOLFGANG VON GOETHE

CERTAIN THINGS IN CONTEMPORARY ZOOLOGY ARE comfortingly consistent. Birds have feathers, and mammals have fur, but reptiles and fish have scales—which is why reports of hairy reptiles and furry fish are decidedly disconcerting!

NEW GUINEA'S ELUSIVE HAIRY LIZARDS

Take, for instance, the seemingly lost hairy lizards of Papua. The following intriguing excerpt appeared in Charles A.W. Monckton's book *Last Days in New Guinea* (1922):

> On the 17th April, 1906, I left Ioma Station...On my way to Mount Albert Edward, I called at a gold working on the Aikora River; a claim there being worked by two men called Bruce and Erickson. Their claim was situated just under one of the main spurs of the mountain; and here I found a funny thing. On the side of the hill the workings had disclosed small caves full of stalactites and stalagmites, and they told me that they had washed out a number of animals or reptiles resembling large lizards with hairy or furry skins. Unfortunately they had preserved no specimens, and I was unable to discover one. I left the miners a small tank of spirits in case any more should be found, as, if the description

given to me was correct, the animal was probably new to science. It was most regrettable that
no more were ever discovered.

Regrettable indeed, but hardly unprecedented, unfortunately,
in the frustrating world of cryptozoology! Assuming that these animals were genuinely reptilian, and not some form of marsupial,
monotreme, or other mammal (and I cannot think of any known
New Guinea mammal that could be readily confused with a
lizard), they would surely have been new to science.

A HIRSUTE VIPER FROM ALGERIA?

So would a hairy snake, and one or two of these have been reported over the years too. Some, of course, were hoaxes, albeit highly
ingenious ones, but others can be less readily discounted. How, for
instance, can the following example be explained? This report
appeared in London's *Observer* newspaper during January 1852:

> In the Algerian paper we read that a hairy viper
> was seen a few days ago near Drariah, coiled
> round a tree. It resembled an enormous caterpillar,
> and was of a brownish-red colour; its length was
> about twenty-two inches. The moment it saw that
> it was observed, it glided into the brushwood, and
> all attempts to discover it were unavailing. The
> authorities of the Museum of Natural History of
> Paris have sent off orders to their agents in Algiers
> to get a specimen of this viper.

Orders or no orders, their agents clearly failed in their appoint-
ed task, because no one seems to have heard anything more about
Algeria's uniquely hirsute vipers. Perhaps it really was nothing
more than an enormous caterpillar, but, if so, I can only presume
that estimates of its size were greatly exaggerated. At least I hope

that they were—any butterfly or moth metamorphosing from a 22-inch-long caterpillar would be a fearsome sight!

BEWARE, BEWARE, THIS FISH WITH HAIR!

As for furry fish: There are at least five different sources of data on file concerning supposed fish sporting a conspicuous covering of hair, fur—or something very like it.

Most mystifying of these is the Japanese hairy fish—a truly bizarre beast whose details were kindly supplied to me by David Heppell, formerly Curator of Molluscs at the Royal Museum of Scotland. According to an early-19th century work, *The World in Miniature: Japan,* edited by Frederic Shoberl, there is said to be a river in Japan that is plentifully populated by a strange amphibious species of creature measuring four to five feet long and possessing a scaly fish-like body but with human-like hair on its head. These animals can apparently come out onto the banks of the river, where they fight or engage in boisterous games with one another, emitting loud cries as they disport in a singularly rowdy, unfish-like manner. However, their rumbustious behavior swiftly transforms into savage aggression if they spy any people, unhesitatingly attacking and killing their hapless human victims by disemboweling them. Yet they do not devour their bodies afterwards. What could have inspired such a strange account?

Assuming that these creatures are more than just a fanciful folktale with no basis in zoological reality, is it possible, as suggested by David Heppell, that they are based upon fur seals? That is to say, could their legend have stemmed from a distorted description or memory of fur seal activity, translocated over generations of retelling from its original marine environment to a freshwater riverine version? Many seemingly impossible beasts of myth and folklore have ultimately been shown to have originated as real animals that were subsequently elaborated by the unsurpassed ingenuity of the human imagination.

In the case of the Japanese hairy fish, we already know that the Pacific waters encompassing Japan are inhabited by the northern fur

FUR SEAL

seal *Callorhinus ursinus*, whose males measure six and a half feet to seven and a half feet long, with females reaching up to five feet. The males possess a hairy head and mane, but with flippers instead of legs they are superficially fish-like, thereby increasing the likelihood that the kind of descriptive liberties routinely employed in dramatic storytelling could readily equip them with that other instantly familiar characteristic of fish—scales.

LODSILUNGUR—HAIRY TROUT OF ICELAND

Scales, conversely, are conspicuous only by their absence from the body of the *lodsilungur*—for this is the legendary hairy trout of Iceland. According to Icelandic mythology, demons and giants specifically create these aberrant fish (often in great quantities, overrunning entire rivers or lakes) in order to punish evil humans, because these creatures are wholly inedible, and are therefore worthless. Needless to say, such a punishment as this would be particularly effective in a country like Iceland, where fishing is of great commercial significance.

Once again, however, such a fanciful fable may have a rather more prosaic core of truth. Perhaps the legend of the *lod-*

silungur is founded upon observations of trout suffering from fungal infections, resulting in an external overgrowth of hair-like mycelia.

A FUR-BEARING FRAUD FROM CANADA

A hairy trout of a very different kind has become one of the most celebrated specimens in the entire zoological collection of the Royal Museum of Scotland. Resplendently garbed in a profuse pelage of white fur, this unique representative of Canada's fish fauna is mounted on a wooden plaque whose label records that it was caught in Lake Superior off Gros Cap, near Sault Ste. Marie, in the district of Algoma, Ontario, and was mounted by a local taxidermist called Ross C. Jobe. According to the label, its dense fur is probably an adaptation to living at great depths, where the prevailing water temperature is exceedingly low.

The Canadian fur-bearing trout is of course a complete fraud, a blatant hoax—but a most successful one. The museum learned of its existence when, already mounted on its informative plaque, it was brought in by its then owner, a lady who had purchased it in good faith while in Canada, and was anxious to discover more about her remarkable exhibit. Once she learned the awful truth, however, that it was merely a common trout to which the furry coat of a white rabbit had been ingeniously attached, she presented it straight away to the museum.

MARCO POLO'S HAIRY FISH

Furry fish #4 has a very esteemed source for its documentation— none other than the famous Venetian traveler Marco Polo. Recording his many travels in China, he reported the sudden drying up of a river and the discovery on the parched riverbed of a strange fish: "fully 100 paces long [whose] whole body was hairy." Moreover, its internal composition may have been as alien as its exterior, for after the people who had found this extraordinary fish had eaten it, they all died.

Perhaps in reality, however, this "hairy fish" was merely some badly decomposed specimen whose "hair" comprised exposed connective tissue, which can yield a startlingly furry appearance when revealed in this manner. And the deaths that occurred after eating it may simply have been due to food poisoning, resulting from the consumption of a rotting fish.

MIRAPINNA—THE "WONDER-FINNED HAIRY ONE"

Unlike the previous quartet, the reality of the fifth type of hairy fish on file is fully accepted by science. Even so, its very existence remained undocumented until as recently as 1956. That was when an amazing new species of fish was formally described from the Azores. Only two and a half inches long, its entire body seemed on first sight to be covered with hair. But when examined more closely, this "hair" was found to be a mass of living body outgrowths containing secretory cells, whose function may be to deter would-be predators.

This remarkable fish was aptly named *Mirapinna esau* ("wonder-finned hairy one"), because not only does it have these curious hair-like outgrowths but also some of its fins' rays are exceptionally long, so that they are more like wings than fins.

A fish with hair is strange enough—but a fish with wings? Now that's another story...

CHAPTER 8

The Sirens of St. Helena

*'Tis true the Manatee and the Dugong are
rather mer-swine than mer-maids; but there
is something in the bluff round head which
may remind a startled observer of the human
form divine.*

P.H. GOSSE—*ROMANCE OF NATURAL HISTORY, SECOND SERIES*

THOSE HIGHLY MODIFIED AQUATIC UNGULATES
(hoofed mammals) the sirenians or sea-cows, represented today by
the manatees and dugongs, are already well known in cryptozoological circles by virtue of the extensively documented (yet incom-
pletely verified) claim that they are responsible for many mermaid
or siren sightings reported from around the world (hence the sea
cows' zoological name, "sirenian"). Other sirenian claims upon the
cryptozoologist's attention include: the possibility that the largest
of all modern-day species, the supposedly extinct Steller's sea cow
Hydrodamalis gigas, still survives; the unmasking in 1985 of the *ri* (an
aquatic mystery beast from New Guinea) as the dugong *Dugong
dugon;* the once disputed existence of the dugong in Chinese
waters; and the likelihood that an unidentified creature reported
from various West African lakes and another such animal from eastern South America's Lake Titicaca may comprise unknown species
of sirenian. In addition, there is the case presented here, one that has
not been documented by other cryptozoologists.

There are three known species of present-day manatee. The
Amazon manatee *Trichechus inunguis* inhabits the estuaries of the
Orinoco and the Amazon; the Caribbean manatee *T. manatus* is

distributed from the coasts of Virginia in the southeastern United States to the West Indies and the northern coasts of Brazil; and the African manatee *T. senegalensis* frequents the coasts and rivers of West Africa from Senegal to Angola. At one time, moreover, there were also persistent reports of putative manatees around the coasts of St. Helena, a small south Atlantic island, almost equidistant from South America and Africa.

MANATEES

In view of the fact that there is a region on the southwestern coast of St. Helena that is actually named Manatee Bay (sometimes spelled "Manati"), one could be forgiven for assuming that there was never any uncertainty about these creatures' identity. In reality, however, this entire matter has still to be resolved satisfactorily, remaining to this day one of the most vexing issues ever raised in relation to sirenian systematics, as evinced by the following selection of reports and opinions.

As documented in a *Proceedings of the Linnean Society of London* article from 1935, Cornish traveler Peter Mundy journeyed in 1655 to India on the *Aleppo Merchant,* and during his return voyage the following year on the same vessel he paid a brief visit to St. Helena. While walking along the beach near Chappell Valley, he saw a strange creature lying ashore and apparently severely injured. Mundy went nearer to examine it:

> However, when I touched it, [it] raised his
> forepart, gaping on mee with his wide and
> terrible jawes. It had the coullor (yellowish) and
> terrible countenance of a lion, with four greatt
> teeth, besides smalle, long, bigge smelling hairs
> or mustaches.

The creature attempted to make its way back to the sea, but Mundy dispatched it with stones. It was evidently very large:

> in length aboutt ten foote and five foote aboutt
> the middle. Some say it was a seale, others notte.
> I terme itt a sealionesse, beeing a femall.

In his journal, Mundy included a sketch of this animal (reproduced in Fraser's account), which leaves no doubt that it was indeed a species of pinniped (seals, sea lions, walruses).

As uncovered by St. Helena resident G.C. Kitching, the Public Records of Jamestown (the island's capital) contain many allusions to alleged manatees or sea cows (including what appears to be the first usage of the name "Manatee Bay," which occurred on January 27, 1679). For example, one such record, for August 28, 1682, listed the capture of "several sea-cows"; and on March 20, 1690, another record noted the following incident:

> Tuesday, Goodwin and Coales brought up for
> killing a Sea-Cow, and not paying the Company's
> Royalty. They desire pardon, and say the Sea-

Cow was very small; the oyle would not amount
to above four or five gallons.

On May 11, 1691, a record mentioned that a sea cow had
appeared on shore at Windward, just a month before traveler
William Dampier visited St. Helena. Dampier became most intrigued
by the alleged existence of manatees around the island's coasts:

> I was also informed that they get Manatee or Sea
> Cows here, which seemed very strange to me.
> Therefore inquiring more strictly into the matter, I
> found the Santa Hellena Manatee to be, by their
> shapes, and manner of lying ashore on the Rocks,
> those Creatures called Sea-lyons: for the Manatee
> never come ashore, neither are they found near
> any rocky Shores, as this Island is, there being no
> feeding for them in such places. Besides, in this
> Island there is no River for them to drink at, tho'
> there is a small Brook runs into the Sea, out of the
> Valley by the Fort.

Returning to the records, on August 29, 1716, they reported
that 400 pounds of ambergris was found in Manatee Bay, and on
September 11, 1739, "A Sea-Cow [was] killed upon Old Woman's
Valley beach, as it was lying asleep, by Warrall and Greentree."

John Barnes's *A Tour Through the Island of St. Helena* (1817) contains a detailed account of these supposed sirenians as described by
reliable observer and St. Helena resident Lieutenant Thomas Leech,
who identified them as sea lions. Yet in complete contrast, another
equally proficient observer, Dr. Walter Henry, just as confidently
identified them as manatees, stating in the second volume of his
Events of a Military Life (1843):

We had sea-cows at St. Helena, the Trichechus
Dugong, but they were not common. When
shooting near Buttermilk Point with another
officer one calm evening, we stumbled on one
lying on a low rock close to the water's edge, and
a hideous ugly brute it was, shaped like a large
calf, with bright green eyes as big as saucers. We
only caught a glimpse of it for a few seconds, for
as soon as it noticed us, it jumped into the sea, in
the most awkward and sprawling manner.

Note that Henry couched his references to these creatures' existence around St. Helena in the past tense. This is because the last recorded appearance of such animals here took place in 1810, when one came ashore at Stone Top Valley beach, and was duly shot by a Mr. Burnham. It measured seven feet long, and 10 gallons of oil were obtained from it. Another of these creatures was also reported in 1810, this time from Manatee Bay.

Since then, St. Helena's purported manatee appears to have been extinct, and as is so often the case it was only then that science began to take an interest in it. After reading an account of this creature in J.C. Melliss's *St. Helena: A Physical, Historical, and Topographical Description of the Island* (1875), in which Melliss claimed that it belonged either to the African or to the Caribbean species of manatee, on June 20, 1899, English zoologist Dr. Richard Lydekker published a short review of the subject in the *Proceedings of the Zoological Society of London,* which contained a number of the accounts given above in this present chapter. Although stating categorically that he did not wish to express a definite opinion concerning whether the animal could truly be some form of sirenian, Lydekker nonetheless ventured to speculate that if this were indeed its identity, it probably constituted a distinct species (perhaps even requiring a separate genus), as he felt unable to believe that it belonged to either of the manatee species nominated by Melliss.

In 1933, the entire matter was the subject of an extensive examination by Dr. Theodor Mortensen of Copenhagen's Zoological Museum, as published in the journal *Videnskabelige Meddelelser fra Dansk Naturhistorisk Forening*. After careful consideration of the varied and often conflicting reports that he had succeeded in gathering, Mortensen came out in support of the views of Mundy and Leech—that the St. Helena manatee was in reality a sea lion. He even identified its species—the Cape sea lion *Arctocephalus antarcticus*—and believed the matter to be closed, reviving it briefly on March 17, 1934, in *Nature* merely to include mention of Dampier's account, which he had not seen when preparing his detailed paper. Certain other records, given in this present chapter but again not seen by Mortensen, were presented as a response to his *Nature* note in Kitching's own *Nature* report, published on July 4, 1936, but Kitching did not express any opinion regarding the creature's identity.

By way of contrast, as outlined within his report of Mundy's sighting, in 1935 F.C. Fraser had leaned very heavily in favor of one specific identity—once again involving a pinniped, but not a sea lion this time. Instead, Fraser nominated a true (i.e. earless) seal—namely, a young male specimen of the southern elephant seal *Mirounga leonina*. As its scientific name suggests, this creature does bear a fancied resemblance to a lion-like beast, and is therefore more reminiscent of a sea lion (albeit one of massive proportions) than are most other true seals. Even so, it bears rather less resemblance to the beast depicted in Mundy's illustration.

Since the 1930s, the St. Helena manatee—or sea lion, or elephant seal—seems to have been forgotten, like so many other "inconvenient" animals, but could it really have been a sirenian? Sadly, the reports on file are not sufficient in themselves to provide an unequivocal answer. All that they can do is offer certain important clues.

For instance, as manatees measure up to 15 feet long the St. Helena beasts were evidently large enough, and their description as calf-shaped by Henry also conforms with that identity. Conversely, the saucer-shaped eyes of Henry's beast conflict markedly with the small, relatively insignificant versions sported by the generally myopic manatees. Large eyes are characteristic of pinnipeds, as are the fearsome jaws and teeth of Mundy's animal. The same can also be said of the latter's moustaches—but as manatees have a bristly upper lip too, this feature is less discriminatory.

If the St. Helena beasts were sirenians, their presence around this island indicates that they may truly have constituted a species in their own right. After all, as Lydekker pointed out in defense of his belief that they belonged neither to the African nor to the Caribbean species of manatee, although it is conceivable that a specimen or two may occasionally be carried from Africa or America to St. Helena, this surely could not occur regularly.

As it happens, there is one notable feature mentioned in a number of the reports cited in this chapter and elsewhere that on first sight greatly decreases the likelihood that these animals belonged to any species of manatee, known or unknown. Although they will rest on the surface of the water in shallow stretches when not feeding, manatees do not generally come ashore. Yet according to several independent accounts, the St. Helena beasts have frequently been seen resting (even sleeping) on the sands or on rocks, completely out of the water, after the fashion of pinnipeds. Also, the large amount of oil obtained from their carcasses is more suggestive of seals than of sirenians.

So are we to conclude that they were not sirenians after all, instead merely large seals or sea lions? Yet if this is indeed all that they were, why did the islanders refer to them so deliberately as manatees or sea cows? It is extremely rare for pinnipeds to be referred to anywhere by such names. In addition, as Lydekker judiciously pointed out, just because *known* sirenians do not normally come ashore voluntarily, this does not mean that there could not be an unknown distinctive species of sirenian that does (or did) come ashore under certain circumstances.

And this is where we must leave the mystery of St. Helena's sirenians-that-might-be-seals: still unsolved, and quite likely to remain that way indefinitely due to the tragic probability that its subject is extinct, lost to science before its identity had even been established.

Finally, there is at least one case on record that constitutes the exact reverse of this one, because it involves some supposed seals that were ultimately revealed to be sirenians. Sea mammals assumed to be seals had been reported from the Red Sea island of Shadwan—but as recorded in 1939 by Paul Budker, when the animals featured in these reports were finally investigated they proved to be dugongs, which are indeed native to the Red Sea.

CHAPTER 9

Giant Mystery Birds

*When I looked towards the sea, I could see
nothing but sky and water; but looking towards
the land, I saw something white; and coming
down from the tree, I took what provisions I had
left, and went towards it. When I came nearer,
I thought it to be a white bowl, of a prodigious
height and extent; and when I came up to it, I
touched it, and found it to be very smooth. I
went round to see if it was open on any side, but
saw it was not; and that there was no climbing
up to the top, it was so smooth. By this time the
sun was ready to set, and all of a sudden the sky
became as dark as if it had been covered with a
thick cloud. I was much astonished at this sudden darkness, but much more so when I found it
occasioned by a bird of monstrous size, that came
flying towards me. I remembered a fowl, called a
roc, and conceived that the great bowl, which I so
much admired, must needs be its egg.*

"Story of Sinbad the Sailor — His Second Voyage," from *The
Arabian Nights' Entertainments: Consisting of a Thousand
and One Stories* (William Milner edition, 1839)

SO ENORMOUS THAT IT WAS VESTED WITH SUFFICIENT strength to carry full-grown elephants aloft in its diabolical talons, the awesome roc is unquestionably among the most famous giant birds of fantasy. But in support of that hoary old adage "Fact is stranger than fiction" are the many examples of gigantic feathered forms from the present and the past whose diverse histories are chronicled within the annals of zoology. Several—such as the ostrich, emu, Andean condor, and the extinct moas of New Zealand—are well known to everyone.

There are also a few famous cryptozoological examples, including North America's thunderbird-related "big bird," and Egypt's heron-like bennu. Both were documented in my book *In Search of Prehistoric Survivors* (1995). There are others, conversely, that will be much less familiar, including the following selection, a veritable phalanx of ambivalent avians whose mighty pinions span the boundaries of reality and reverie, submitting for investigation some of the world's greatest crypto-ornithological challenges.

AUSTRALIA'S "ABOMINABLE SPINIFEX MAN"

The ratites comprise a loose taxonomic group of giant flightless birds lacking a keel on their sternum (breastbone), and include the ostrich, emu, cassowaries, and rheas, as well as a number of extinct types, such as the New Zealand moas, the Madagascan elephant birds, and the Australian mihirungs. Several unexplained anomalies of ornithology appear to involve ratites, but one of the most unexpected must surely be this first example.

During summer 1970, dingo hunter Peter Muir found and photographed some strange two-toed tracks in the spinifex (a spiny-seed grass) desert area near Laverton, Western Australia. Local aboriginals claimed that they were from an ogre-like monster, the *tjangara* or spinifex man. Scientists initially assumed that they were merely emu tracks, but swiftly retracted this view because emus leave *three*-toed versions. Only one known creature produces two-toed tracks like those at Laverton—*Struthio camelus*, the African ostrich!

Yet the concept of ostriches living wild in Australia is by no means ludicrous. Far from it. Small populations of feral (run-wild) ostriches still persist north of Adelaide, South Australia—descended from specimens released by ostrich farmers after World War I, when the feather market collapsed.

As the ostrich is a desert-hardy bird that can travel great distances in short periods of time, in 1971 zoologist Ivan T. Sanderson suggested that some ex-farm specimens may not only have survived and bred but also have discreetly extended their range across the intervening desertlands into Western Australia, and thence to Laverton. (Incidentally, it is known that ostriches were released in Western Australia before 1912, but these "officially" died out without establishing a population.) This would offer a plausible explanation for the mystery of the two-toed tracks and their unseen originator(s)—aptly dubbed by Australian wags "the abominable spinifex man." Indeed, if the Laverton environs were not so sparsely populated, the existence of ostriches there may have been confirmed by now.

QUEENSLAND MOAS AND LIVING MOAS?

It is well established that the extinct moas comprised a group of ratites exclusive to New Zealand—which is why the little-known tale of the Queensland moa is worth retelling. In 1884, this zoogeographical heretic was christened *Dinornis queenslandiae* by C.W. de Vis (at that time the director of the Queensland Museum), who based its species on a fossilized, incomplete left femur, spotted by him in a collection of bones from King's Creek, Queensland. Naturally, this specimen attracted great interest among ornithologists, as it extended the moas' distribution very considerably. No longer were they a novelty of New Zealand—always assuming, of course, that it really was a moa.

Over the years, this assumption became a much-debated issue. In 1893, for instance, F.W. Hutton deemed the bone to be from a cassowary-like species that probably represented the common ancestor of emus and cassowaries (which are currently classed together within the same taxonomic order). Then in 1949 its species was readmitted to the moa brotherhood, when Dr. W.R.B. Oliver of Wellington's Dominion Museum renamed it *Pachyornis queenslandiae*. By the early 1960s, conversely, it was back among the emus and cassowaries, when in 1963 Alden H. Miller classified it as an emu, dubbing it *Dromiceius (=Dromaius) queenslandiae*. It seemed as if this contentious species would be spending the rest of time ricocheting from one ratite family to another—but then came the study that finally brought its taxonomic tribulations to a long-awaited end.

This was when, in April 1967, osteologist Ron Scarlett from New Zealand's Canterbury Museum not only examined this femur and revealed that it was indeed from a *Pachyornis moa*, but also paid close attention to the other bones in the original King's Creek collection within which it had been found by de Vis back in the 1880s. In so doing, Scarlett recognized that the femur was strikingly different in general appearance and color from the rest of this collection's material, and clearly had not been obtained with it. In addition, his very appreciable experience with moa bones derived from caves, Maori middens, swamps, and other sources of such remains enabled him to reveal something even more significant— the femur was readily identifiable as a bone originating from a midden in New Zealand's South Island (*Memoirs of the Queensland Museum*, 1969). It was not from Australia at all!

The famous Queensland moa was just another non-existent creature that had been granted a transient reality by the evocation of inaccurate information and incomplete investigation—a mere monster of misidentification, nothing more. *Pachyornis queenslandiae*, R.I.P.!

As for the moas in their native New Zealand, most ornithologists believe that these have been resting in peace for several centuries—with possibly a couple of notable exceptions. Although some of the 11 species of moa were six feet to 12 feet tall, and ranged in appearance from the sturdy and elephantine to the graceful and ostrich-like, a few were relatively small and inconspicuous. One of these latter species, the upland moa *Megalapteryx didinus,* is of particular interest because it may have survived until as recently as the mid-1800s, and perhaps even later than that. Some ornithologists have expressed a degree of optimism that it could have survived undetected into the present day.

As discussed in my book *Extraordinary Animals Worldwide* (1991), this hope is fuelled by reports concerning a mysterious "giant kiwi," said by the Maoris to have spurs on its feet and to be the size of a turkey. According to some writers, it is known locally as the *roaroa,* but in fact this is the name given by the Maoris to the largest *known* kiwi, the great-spotted kiwi *Apteryx haasti.* Nevertheless, *A. haasti* is much smaller than a turkey and does not have spurred feet. Nor did the upland moa, but it was indeed turkey-sized. Moreover, it is very likely that in life it closely resembled a very large kiwi (*Megalapteryx* translates as "big kiwi"). Also, it had a well-developed clawed hallux on each foot that might conceivably be likened or referred to by non-specialists as a spur (so too do kiwis, but they are of course smaller birds than the upland moa). So is the elusive "giant kiwi" really a miniature moa?

After utilizing the latest computer technology in conjunction with fossil specimens (revealing its throat structure) to reconstruct the upland moa's voice, in February and March 1978 a Japanese expedition led by Gunma University biologist Dr. Shoichi Hollie trekked through South Island's Fjordland, playing the reconstituted cry in the hope of enticing a living *Megalapteryx* into view. Their call went unanswered.

When is a moa not a moa? When it is actually an antler-bearing stag! That, at least, is one official view held in relation to photos obtained of what its stunned observers have claimed to be a living moa of the six-foot-tall variety, far removed indeed from the diminutive *Megalapteryx.* This extraordinary episode in the continuing controversy regarding the prospect of moa survival hit the news headlines worldwide in late January 1993.

According to the accounts, of which the most detailed were those of New Zealand's own *Christchurch Press* newspaper (which gave the unfolding drama front-page coverage for several days), the bird was allegedly seen shortly after 11:00 a.m. on January 20, 1993, by three experienced hikers tramping in Canterbury's Craigieburn Range, South Island. The trio consisted of Paddy Freaney (a former instructor with the British Army's highly renowned Special Air Service, i.e. the S.A.S., who now runs Bealey Hotel at Arthur's Pass), Rochelle Rafferty (a gardener employed at Freaney's hotel), and Sam Waby (head of the art department at Aranui High School). After hiking for approximately four hours, they had reached a Harper Valley riverbed. Waby paused to take a drink when suddenly Freaney caught sight of an extremely large bird standing by a bush, roughly 40 yards away. Hardly believing his eyes, he whispered to the others to look, and as soon as Waby saw it he unhesitatingly identified it as a moa.

Freaney, who was closest to the bird, stated that its body height was about three feet, and that it possessed a long thin neck that was also about three feet, which terminated in a small head and beak. Its legs were very large and thick, clothed in feathers almost down to the knees, but the lower portions were bare, as were its huge feet. Its body's feathers were reddish-brown and grey. As soon as it saw its three observers, the bird ran away, racing across a small stream as Freaney vainly attempted to keep pace with it in order to take some photographs. At a distance of 35 to 40 yards, he succeeded in snapping one blurred photo of the bird itself, by bushes and rock formations, as well as some of what he thought might be a wet footprint left behind by it on a rock and others in shingle by the riverbed. The bird vanished from sight amid some dense bushes. The entire incident, from Freaney's initial sighting of the bird to its successful disappearance, lasted no more than 30 seconds, but the three observers were totally convinced that it was a genuine moa.

When news of this incredible occurrence reached officialdom, the reaction elicited was a turbulent blend of bewilderment, caution, excitement, suspicion, and outright astonishment. Discounting the possibility of a hoax, Dr. Ken Hughey, Protection Officer in New Zealand's Department of Conservation (DOC), announced that the creature in the photograph was either a bona fide moa or an escapee emu—for as stated by Andy Grant, the DOC's Protected Species Officer, it was clearly not an ostrich, the only other bird of comparable form.

As it happens, there is an emu-breeding center in Canterbury, at the Isaac Wildlife Trust, but inquiries revealed that none of its birds had escaped. Similarly, although 1992 had seen the Trust's first sales of emu chicks, these had not taken place until October and November and the chicks had only been six weeks old. Even if one of these had escaped from its new home, it would have still been far too small to explain the Freaney trio's encounter.

And so, on January 26 the DOC announced that it would be sending an investigative team to the locality of their sighting, in search of fecal pellets, feathers, and any other evidence for the presence of what would unquestionably comprise—if its existence were indeed formally confirmed—the greatest zoological discovery in New Zealand in 200 years. And then it all went sadly wrong. On that same day, Moana Railway Station's owner, Dennis Dunbar, revealed in a *Christchurch Press* report by Diane Keenan that he and Freaney had become friendly rivals in striving to obtain the best stories for media coverage, and that Freaney had told him a few months ago that his next feat would outdo anything that had previously taken place between them.

By the following day, the DOC had announced that it would not be sending out its team after all. Dr. Hughey revealed that in addition to the disclosure of Dunbar's comments by the media, a number of rumors lending weight to the possibility that the sighting had been a hoax had been reported to the DOC—hence it would be making further contact with the trio of observers before making any final decision regarding a formal search.

This was followed on January 27 by Dunbar's retraction of his earlier statement, and by Freaney's announcement that he would personally mount a search for evidence to substantiate the sighting, although he would still be only too happy to offer as much information to the DOC as he could. While all of the deliberations regarding the nature of the sighting were taking place, however, the all-important photograph of the creature itself was being subjected to three days' worth of computer enhancement analysis by image processing specialists in Canterbury University's Department of Electrical and Electronic Engineering. And on January 29, one of these researchers, postgraduate student Kevin Taylor, announced that in his opinion the resulting "de-blurred" version of the photo showed a large bird. Conversely, after examining this, paleoecologist Dr. Richard Holdaway stated that the creature's neck was too thick to be that of a bird, and that he instead believed it to be a male red deer.

Initially, it may well seem inconceivable that any subject depicted in a photo could be variously identified as creatures so dissimilar from one another as a tall two-legged bird and an antler-bearing four-legged stag! However, as readers had already discovered for themselves when it was reproduced in color by the *Christchurch Press* in its front-page coverage on January 25, the photo's portrayal of the creature is anything but straightforward. All that can be readily perceived of it is a thickset horizontal body, brown in color, and a sturdy upright neck, bearing what appears to be a head at its uppermost point. On top of, or directly above, the head are some dark horizontal objects, which can be interpreted with an equal degree of success as antlers, branches from the bushes sited immediately to the left of the creature, shadows, or rock formations. As for that most crucial of features—the creature's number of legs—this is not visible, because the lower portion of its body is completely hidden behind rocks, an obstacle that not even computer enhancement can counter. And the photos allegedly depicting the creature's wet footprint gained short shrift from both Dr. Holdaway and Canterbury Museum's moa expert Beverly McCulloch, as they deemed it unlikely that this was the footprint of any type of bird.

Nevertheless, Kevin Taylor remained interested in pursuing the investigation via his own computer enhancement techniques, but this time by directly utilizing the original negatives of Freaney's photographs. This was permitted by Freaney in early February, but the results were not conclusive. A few days afterwards, Freaney revealed that he would be sending them for computer enhancement to an English university astrophysicist, and had received the promise of part-sponsorship from an anonymous businessman for a planned search for the bird itself later that year. In the meantime, however, some very startling—and extremely unexpected—support for the Freaney trio's moa claim came to light.

During the second week of February, hiker Bruce Spittle and his son Malcolm Moncrief-Spittle, visiting the Cragieburn area, had stayed at its Bealey Hut accommodation, and while reading through some of the entries in the hut's intentions (visitors) book had spotted a remarkable entry penned by two German hikers during the previous year. Dated May 19, 1992, it revealed that Franz Christianssen and Hulga Umbreit had spent two days visiting the Cass-Lagoon Saddle area of Harper Valley, and whilst there had been "very surprised to see two moas"! The validity of the entry's date was duly confirmed by Geoff Goodhew, a Christchurch hiker who had been a member of the party that had made use of the hut just three days after the two Germans. This meant that the Freaney encounter now had a notable precedent—an encounter with alleged moas made in precisely the same location, but eight months *earlier*. Needless to say, Freaney and his two colleagues were delighted and greatly encouraged by this apparent vindication of their claim, but the DOC remained unimpressed, discounting the German report as insufficient evidence for mounting a moa quest.

And there the matter remains for the present time. Nothing further of note has emerged regarding this latest in a long line of reputed moa resurrections—and quite probably nothing will. For myself, although I feel comfortable in accepting the possible existence undetected by science of a surviving species of modest-sized

Megalapteryx moa in certain of the lesser-explored corners of South Island, the prospect of a viable population of mighty six-foot-tall moas eluding detection in a region as regularly traversed by man as the Craigieburn Range is not one that inspires me with a notable degree of optimism. Of course, scientists were expressing much the same views concerning the brilliantly plumaged takahe *Porphyrio (=Notornis) mantelli,* a flightless relative of the moorhens and gallinules believed extinct since 1899—until it was sensationally rediscovered in December 1948. To be continued…?

A BEWILDERMENT OF DWARF EMUS

Nowadays, the familiar emu *Dromaius novaehollandiae* of mainland Australia is the only type of emu still in existence, but not so long ago it had three dwarf relatives.

Just north of Tasmania is the Bass Strait, containing a small islet known as King Island; and just to the south of Adelaide, South Australia, is another one, called Kangaroo Island. Each one harbored its very own, distinctive species of emu—both of which had been discovered by the Western world in the 19th century's opening years, only to vanish utterly and irrevocably by the 1830s. Since then, the histories of these two scarcely known, lost species have been inextricably intertwined by a Gordian degree of confusion and contradiction—which only now, in the following account, is finally unraveled.

In 1802, Kangaroo Island was briefly visited by a French naval expedition to Australia, led by Captain Nicholas Baudin, and some living specimens of its endemic emu were taken on board. At least three survived the arduous journey back to France, and two were forwarded to the Empress Josephine; the third survived for a time at Paris's Jardin des Plantes.

There is no firm record concerning the fate of the first two specimens' remains after their deaths in 1822—events that most probably marked the extinction of this entire species, because it had died out by then on Kangaroo Island itself too, a consequence of devastating brush fires and hunting. In contrast, the skin of the Jardin des Plantes' individual was preserved, and survives today in mounted form at Paris's Natural History Museum. The specimen upon which the Kangaroo Island emu's scientific description has traditionally been based, this bird had much darker plumage than the Australian emu, alleviated only by its pale face, throat, and back, and was somewhat shorter too—so that its species also became known as the black emu and as the dwarf emu. Several different scientific names have been assigned to it as well, including *Dromaius diemenianus* and *D. ater*, but in terms of confusion these nomenclatural inconsistencies were as nothing compared to the taxonomic turmoil that would ultimately ensue concerning this enigmatic emu.

Within his impressive *Rare and Vanishing Australian Birds* (1978), Peter Sclater pointed out that during its Antipodean travels of 1802 the Baudin-led naval expedition had collected emus not only from Kangaroo Island but also from King Island. This disclosure greatly reduced the significance of the preserved skin from the Jardin des Plantes' emu, because there are no records that categorically identify from which of these two islands this all-important, unique ornithological specimen had actually been obtained. Consequently, no one could be absolutely certain any longer that it really had been a Kangaroo Island emu. After all, who could guarantee that it was not a King Island emu?

Based as it was, therefore, upon a skin that may not even have been derived from a representative of its own species, the Kangaroo Island emu was facing an acute identity crisis. Its King Island relative, meanwhile, had fortunately avoided this same fate (though not the fate of extinction—by the 1830s it too had been wiped out by a combination of forest fires and hunting). That was because this species had been based upon collections of fossilized emu bones (unearthed from shoreline sand dunes) whose precise localities upon King Island had been fully documented and authenticated within the scientific literature.

Accordingly, in the early 1980s Dr. Shane Parker of the South

Australian Museum decided that the best way to deal with the confusion surrounding the Kangaroo Island emu would be to pursue a similar course of action—by rejecting outright the preserved skin as a reliable representative of this species, and by discounting all of the scientific names that the skin had inspired. In their place, he provided a brand-new description of this near-forgotten ratite (*B.B.O.C.*, 1984), based solely upon the structural characteristics of a series of emu bones whose collection upon Kangaroo Island had been precisely recorded by the scientists who had obtained them there. These revealed that this controversial bird had been slightly taller than King Island's but still much shorter than the surviving emu of mainland Australia.

As for the Kangaroo Island emu's scientific name, Parker formally changed this to *D. baudinianus* (commemorating Captain Baudin). This new name is particularly welcome, because its two principal predecessors, *D. ater* and *D. diemenianus*, had been responsible for inciting even further chaos concerning this already much-muddled bird. In earlier days, for example, the name *D. ater* had also been applied by some writers to the King Island emu (otherwise referred to as *D. minor*); following on from Parker's paper (and in many other publications since then), this name has now been formally assigned exclusively to the King Island species (thereby replacing *D. minor* as its official scientific name). As for *D. diemenianus*, this is all-too-readily mistaken for *diemenensis*, the subspecific name of the Tasmanian emu, yet another lost form.

Even shorter than the King Island emu, and lacking the Australian species' characteristic black feathers on the upper reaches of its neck, the Tasmanian emu is known in full as *D. novaehollandiae diemenensis*, and is represented today only by bones and a few complete preserved specimens in various of the world's major museums. As already mentioned, many ornithologists believe that the emus of Kangaroo Island and King Island were depleted by brush fires, sweeping unchecked across their sea-encircled homes; similar events may have contributed to the decline of their

Tasmanian relative too, but there is a rather more insidious factor that also played a notable part in this particular emu's passing.

Like those on Australia, the emus of Tasmania were greatly prized as game by the gun-toting fraternity present among Western settlers during the early 19th century. But as there were far fewer emus on Tasmania than on the mainland, *D. n. diemenensis* rapidly became very rare. In order, therefore, to provide greater numbers for shooting, emus from Australia were released in Tasmania, which resulted in widespread interbreeding between this island's native emus and those from Australia. Inevitably, it was not long before virtually all of the emus on Tasmania were either hybrids or pure-bred Australian specimens—by 1850, if not earlier, the last true Tasmanian emus were gone.

During the 1920s eight Australian emus were released onto Kangaroo Island, the first emus seen here since its own species had died out almost a century before. A few of these interlopers from the mainland were still present in the 1980s.

AUSTRALIA'S GIANT EGG AND MADAGASCAR'S ELEPHANT BIRDS—A CRYPTIC CONNECTION?

One of the most controversial ornithological mysteries from Down Under concerns the origin of a gigantic egg that was discovered by rancher's son Vic Roberts and his friend Chris Morris at Nannup, in the Scott River region of Western Australia's southernmost tip, in 1930, ensconced among some sand dunes about 500 yards from the sea. An emu, or even a feral ostrich, might seem a feasible expla-nation for it, until we consider that its diameter was only margin-ally under one foot, i.e. almost twice that of an average-sized ostrich egg, which is the biggest egg laid by any bird alive today! The other dimensions of this oological monster are equally astonishing—a width of eight and a half inches, a circumference of roughly 26 and a half inches, and a capacity of almost one and a half gallons (i.e. equivalent to about 135 hens' eggs!). It was eventually presented by Roberts on permanent loan to the Western Australian Museum, where it still resides today, but the identity of the gargantuan bird that must have laid it has puzzled the zoological world for decades.

In 1962, the egg was examined by the widely known Australian naturalist Harry Butler, who accurately pointed out in a later report (*Science Digest*, March 1969) that it most closely compared in size with the enormous eggs produced by the now-extinct aepyornids or elephant birds—from Madagascar!

Comprising a family of uncommonly robust ratites, most species of elephant bird were native to Madagascar or to mainland Africa. Some unexpected proof that the Canary Islands were connected to the African land mass until as recently as 12 million years ago was disclosed by German scientists Drs. E.G. Franz Sauer and Peter Rothe in a *Science* article for April 7, 1972, when they announced that fossil elephant bird (and also ostrich) eggshells dating from the early Pliocene had been found on the island of Lanzarote. Yet despite their relatively wide distribution, all but one species had died out in prehistoric times.

The lone exception, however, was indeed exceptional—for it was the largest elephant bird of all. Appropriately named *Aepyornis maximus*, the great elephant bird, this enormous creature stood more than 10 feet tall and weighed almost 1,000 pounds. It is very likely that it was still alive as recently as the 17th century, because within his *Histoire de la Grande Isle de Madagascar* (1658), Admiral Étienne de Flacourt referred to a giant bird that he called the *vouroupatra* (possibly a misspelling of *vorompatra*), which was still occasionally reported at that time by the Ampatres (Antandroy) tribe of southern Madagascar. According to de Flacourt, it laid huge eggs like the ostrich's, and in order to avoid being captured it only inhabited those localities least frequented by man.

Tragically, the *vouroupatra* was not destined to survive for much longer. Extensive deforestation and the attendant disappearance of much of the wooded swampland terrain constituting this huge but highly reclusive bird's habitat swiftly brought to an end the empire of the elephant birds, summarily consigning

Aepyornis maximus—the last, and greatest, of its line—to the permanent exile of extinction.

By the beginning of the 18th century, the *vouroupatra* had gone, and for many years afterwards it would be deemed by Western science to be as imaginary as the mighty roc of Arabian Nights fame—a supremely ironic development, considering that most ornithologists nowadays agree that the legend of the roc and its enormous eggs was almost certainly inspired by the former existence of the great elephant bird.

Today, of course, the reality of the *vouroupatra* is no longer in doubt, endorsed by the silent testimony in museums worldwide of its towering skeletons—and its colossal eggs. They have been frequently dug up in fully preserved subfossil state from the coastal sands of this species' erstwhile island dominion ever since the first three to be seen by Western scientists were brought back from southwestern Madagascar to Paris in 1851 by a French sea captain called Abadie. Their remains have accorded *Aepyornis maximus* a lasting place in the zoological record books, because they are the biggest eggs ever documented from any creature known to have existed on Earth. With a fluid capacity often exceeding two gallons, they are greater than ostrich eggs and even those of the biggest dinosaurs.

It is little wonder, therefore, that Butler favored the identification of Roberts's titanic egg as that of an elephant bird—but if this *is* its correct identity, how can we explain its discovery a full 4,000 miles beyond Madagascar, on a sand dune in Australia? In answer to this, Butler postulated that the egg may simply have drifted there, borne upon the prevailing ocean currents.

This hypothesis is far from being as implausible as it might initially appear, for it so happens that the Indian Ocean and the Southern Ocean converge at the very coastal point where the egg was found, so that a great amount of flotsam and jetsam is frequently washed ashore there. In conjunction with this, Butler also noted that during the early period of Australia's settlement by Westerners, French sealers and whalers operating in the coastal region containing the egg's provenance generally used Madagascar as a base. As elephant bird eggs were (and still are) greatly valued novelties here, it is quite reasonable to assume that some of those French seagoers traveling to Australia from Madagascar may have brought back such eggs as souvenirs (as Abadie did)—and if any happened to be lost overboard (during a storm or shipwreck, for instance), they would certainly float ashore.

This, then, is one solution to the mystery of Australia's giant egg.

THE MARVELOUS MIHIRUNGS — WHEN DID THEY REALLY DIE OUT?

Even so, an exceedingly out-of-place *Aepyornis* egg was not the only identity proposed by Butler for it. He also suggested that it may have originated from one of the huge species of extinct, superficially emu-like birds whose fossil remains have been found in parts of Queensland and South Australia, some of which date from as recently as the late Pleistocene.

These were the dromornids, also known as the mihirungs, and comprising a family that includes among its members the most massive, heaviest bird of all time—outweighing even the hefty *Aepyornis maximus*. Tipping the scales at half a ton, and perhaps as tall as 10 feet, this plant-eating behemoth of a bird is Stirton's mihirung *Dromornis stirtoni*, whose thunderous footsteps echoed through the forests of northern Australia two and a half million years ago. Hardly a creature likely to be overlooked by science, one would have thought—which makes it all the more paradoxical that the mihirungs are not only among the largest but also among the least known of all giant birds.

Indeed, they scarcely received a mention in ornithological works until the late 1970s, when they belatedly began to generate interest and recognition by virtue of the extensive mihirung research being carried out by palaeontologist Dr. Pat Rich, from Melbourne's Monash University, who has presented her comprehensive findings in numerous publications. These include two important multi-author volumes—*Kadimakara: Extinct Vertebrates of Australia* (1985), and *The Antipodean Ark* (1987).

As almost all of today's burgeoning wealth of information on these enormous birds has thus been made available only in the last 15 years or so, one might be forgiven for assuming that science has not known of the mihirungs for very long, but this would be totally incorrect. In fact, the first evidence for their former existence had been brought to scientific attention as long ago as 1830. This was when Major (later Sir) Thomas Mitchell obtained a single mihirung femur bone during a fossil-collecting visit to the Wellington Valley caves of western New South Wales. Since then, many other fossilized remains have been found, so that several different species are now on record. Initially thought to have been giant emus (and occasionally termed "super-emus"), but built more on the lines of the much sturdier elephant birds in some cases (such as *D. stirtoni*), mihirungs are now held to be more closely related to waterfowl (ducks, geese, and swans) than to emus or other ratites, and have even been nicknamed the demon ducks of doom!

Mihirungs were once thought to have died out long before humans arrived in Australia, but this no longer appears to be true. Their last representative was Newton's mihirung *Genyornis newtoni*, first made known to science in 1896 and currently known from fossils dating from as recently as 15,000 years ago. As humans are believed to have reached this island continent at least 40,000 years ago, humans and mihirungs co-existed for 25,000 years or more. In addition, memories of *Genyornis*—believed to have stood seven feet tall and endowed with a noticeably deep beak—seem to have been retained in the oral traditions of certain aboriginal tribes. In western Australia, for example, the Tjapwurong tribe's legends mention a giant bird that greatly resembles the probable appearance in life of *Genyornis* as deduced from its skeletal anatomy. They refer to this bird as the *mihirung paringmal*—from which "mihirung" was later borrowed by science to provide a vernacular name for all members of the dromornid family. *Genyornis*-like birds also occur in examples of aboriginal rock art from northern Queensland.

As these legends are unlikely to have been preserved for as long as 15,000 years, and as the paintings are of much more recent date than this too, if the birds featured in them truly are mihirungs then this indicates that *Genyornis* persisted to a later date than currently provided by fossil evidence alone. What a pity that this spectacular bird did not survive right up to the present day, to be seen alive by modern man. But perhaps it has! According to a report in *Fabuleux Oiseaux* (1980) by Jean-Jacques Barloy and Pierre Civet, at least one such encounter may have taken place very recently.

In 1967, George and Jean Rollo were exploring the dense forest near Robertson, on the southern coast of New South Wales, when they were startled by a sudden loud noise, resembling the cracking of branches being pushed roughly to one side. Turning round, they were amazed to see, at close range, a gigantic bird that they likened to some sort of "super-emu," and which they estimated to be more than nine feet tall, with a long neck but only vestigial wings and fairly sparse plumage. Before they were able to make any further observations of its appearance, however, the bird charged towards them, so that they were forced to withdraw with all speed! As the true Australian emu *Dromaius novaehollandiae* rarely exceeds six feet in total height, the Rollos' antagonist clearly did not belong to this species; nor did it resemble an ostrich, thereby eliminating wide-ranging feral specimens from further consideration too. The only other ratite known to exist in Australia today is the double-wattled cassowary *Casuarius casuarius*, a mere five and a half feet tall. And so Barloy and Civet cautiously offered a mihirung as an identity.

Needless to say, this is a very remote one, but it does provide a thought-provoking tie-in with another ratite-related mystery Down Under. The fossils of *Genyornis* show that, prior to its extinction, its three-toed feet had been evolving toward a two-toed state, paralleling that of the ostrich. If *Genyornis* had survived into the present day, it is possible, therefore, that continued evolution would have completed this transformation to yield a genuinely two-toed species of mihirung. Is it just coincidence that the tracks of such a bird would have corresponded perfectly with those of Australia's "abominable spinifex man"?

And speaking of ratite riddles, there is still the Western Australian Museum's unidentified giant egg to consider. The short-lived flicker of scientific interest ignited by Harry Butler's *Science Digest* report focused upon his aepyornid proposal to the absolute exclusion of his less-publicized, alternative hypothesis — that the egg was that of a mihirung. During the early 1980s, however, an oological discovery was announced that seemed on first sight to offer support for this latter suggestion.

In 1981, biologist Dr. David Williams from the University of South Australia revealed that several hundred fragments of enormous fossil eggshells had been collected from an eroding coastal sand dune near South Australia's Port Augusta (*Alcheringa*, 1981). Radiocarbon dating indicated that they were at least 40,000 years old, and morphological comparisons implied that they were probably from *Genyornis* eggs. Based upon the fragments' dimensions, estimates of the likely size of a complete *Genyornis* egg gave a length of roughly six inches, a width of five inches, and a weight of 2.86 pounds, thereby comparing with the very largest ostrich eggs on record. Moreover, as there is no reason to suppose that these particular fragments originated from unusually large *Genyornis* eggs, it is possible that this bird produced even bigger ones.

Taking all of this together, and supplying the additional fact that Port Augusta shares the coastline serving Nannup, site of the giant mystery egg's discovery by Vic Roberts, surely the latter egg must be a mihirung's? Although superficially persuasive, in reality this identity is far less convincing than it may initially appear. This is because water, percolating through the dune's sand while the egg lay there for several millennia (the scenario as predicted by the mihirung hypothesis), would have completely dissolved the shell. Fragments may have survived, but not an entire egg. Also, as I learned from Dr. Ron Johnstone of the Western Australian Museum's Department of Ornithology, despite rumors to the contrary the egg is *not* fossilized, but is of geologically Recent date (with a very porous surface, heavily pitted by sand-blasting), renewing support for an aepyornid origin (unless it was from an undiscovered modern-day mihirung?!).

I am greatly indebted to Dr. Johnstone for bringing to my attention a second, very comparable case of a dramatically out-of-place egg—one of much later date but of great relevance to this longstanding mystery. In January 1991, Ongerup farmer Kelly O'Neill discovered a 4.4-inch-long, 3.2-inch-wide, barnacle-encrusted egg washed ashore near Bremer Bay, on Australia's South Coast and later publicized in the *Great Southern Herald* (June 19, 1991, issue). It was identified at the Western Australian Museum as that of a king penguin *Aptenodytes patagonica*, a species native to Kerguelen and other sub-Antarctic islands sited off the southeastern coast of Africa! This is not an isolated case either—17 years earlier, ornithologist Geoff Lodge had found a similar specimen…washed ashore close to Nannup!

As Dr. Johnstone states, this provides additional support for Butler's belief that large eggs can be transported by ocean currents for thousands of miles from their place of origin to the South Coast of Australia, and greatly strengthens his theory of an aepyornid identity for Australia's anomalous giant egg—a theory to which Johnstone also subscribes. After more than 60 years it would seem, therefore, that the mystery of one unexpected egg's arrival in Australia has at last been solved—by the arrival here of another one!

But even this is not the end of the story. In March 1993, three children—Jamie Andrich (9), Kelly Pew (8), and Michelle Pew (6)—dug up a massive egg, with a circumference of 32.2 inches, out of some sandhills about half a mile away from the beach at Cervantes, 150 miles north of Perth. Dr. Ken McNamara, the Western Australian Museum's senior paleontological curator, speculated that it was either another *Aepyornis* egg that had been carried by ocean currents from Madagascar, or, of even greater scientific importance, the first fully intact mihirung egg recorded by science. It was later shown to be the former.

A CONFUSION OF CASSOWARIES

Most closely related to the emu among living ratites and distributed widely through Australasia, the cassowaries constitute a trio of imposing, forest-dwelling species with black, spine-like feathers enlivened by brilliantly colored patches of red, mauve, or blue skin on their neck; a multifarious assemblage of commensurately gaudy wattles; and a bony helmet-like casque, again of varying appearance. The native tribes sharing their jungle domain often keep young cassowaries as pets, but greatly fear the adult birds on account of their formidable claws—with which, the natives aver, they can readily disembowel with a single kick anyone foolish enough to threaten them. Even so, this does not prevent many tribes from utilizing cassowaries as a form of feathered currency, trading living specimens or select portions of dead ones (particularly the casque, claws, and feathers) far and wide in exchange for useful items such as domestic livestock—and wives! The late Dr. Thomas Gilliard, an expert on New Guinea avifauna, learned that in Papua the rate of exchange for one live cassowary is eight pigs, or one woman!

During the early 17th century, scholar Charles de l'Écluse placed on record the eventful history of the first cassowary ever seen in Europe — a much-traveled specimen originally captured on the Moluccan island of Seram (Ceram), but brought back to Amsterdam in 1597 from Banda (another Moluccan island) by the first Dutch expedition to the East Indies. Given to the expedition by the ruler of the Javanese town of Lydajo, it lost no time in becoming a much-coveted cassowary following its arrival in Holland. Effortlessly ascending ever higher through the rarefied strata of European high society, after a period of several months as the star of a highly successful public exhibition at Amsterdam, this distinguished bird passed into the hands of the Count of Solms and journeyed to the

Hague. Later it was owned for a time by the Elector Palatine, before attaining the zenith of its fame by becoming the property of no less a personage than Emperor Rudolph II of the Holy Roman Empire.

Its species became known as *Casuarius casuarius,* and is the tallest of the three modern-day species of cassowary, averaging five and a half feet in height (including its lofty casque), and generally sporting two wattles on its neck, so that it is most commonly called the double-wattled cassowary. Based in most cases upon only the most trivial of differences in the color, number, and shape of these wattles, a bewildering plethora of species all supposedly distinct from *C. casuarius* were described during the 19th century. These included *C. australis* (from northeastern Australia), *C. beccarii* and *C. bicarunculatus* (both from the Aru Islands), *C. tricarunculatus* (Geelvink Bay), and *C. unappendiculatus* (Salawati Island).

Only the last-mentioned of these, however, is still recognized as a genuinely separate species. Standing four and a half feet to five feet tall and known as the single-wattled cassowary, *C. unappendiculatus* also inhabits New Guinea and the offshore islands of Misol and Japen. It was first made known to science by Edward Blyth in 1860, from a specimen brought back alive to Calcutta from an unrecorded locality.

Three years earlier, Dr. George Bennett, a surgeon and biologist from New South Wales, had recorded the existence on New Britain of a cassowary whose unusually small size, lack of wattles, and noticeably flattened casque left no room for doubt that, unlike so many other forms being described and named, this really was a radically new, well-delineated species. Known to the natives as the *mooruk* and only three and a half feet tall, it was later christened *C. bennetti,* Bennett's cassowary, by John Gould. Greatly intrigued by this dwarf species, Bennett obtained two *mooruks* from New Britain and gave them the freedom of his home—discovering that they made entertaining if inquisitive house guests, as succinctly summarized in W. H. Davenport Adams's *The Bird World* (1885):

The birds...were very tame; they ran freely about his house and garden—fearlessly approaching any person who was in the habit of feeding them. After a while they grew so bold as to disturb the servants while at work; they entered the open doors, followed the inmates step by step, pried and peered into every corner of the kitchen, leaped upon the chairs and tables, flocked round the busy and bountiful cook. If an attempt were made to catch them, they immediately took to flight, hid under or among the furniture, and lustily defended themselves with beak and claw. But as soon as they were left alone they returned, of their own accord, to their accustomed place. If a servant-maid endeavoured to drive them away, they struck her and rent her garments. They would penetrate into the stables among the horses, and eat with them, quite sociably, out of the rack. Frequently they pushed open the door of Bennett's study, walked all around it gravely and quietly, examined every article, and returned as noiselessly as they came.

The discovery of Bennett's cassowary was followed by the documentation of other, similarly undersized, wattle-less types, initially treated as distinct species. These included Westermann's cassowary C. *westermanni* (Jobi Island) and the painted cassowary C. *picticollis* (southeastern New Guinea), but they are all conspecific with C. *bennetti*.

Among the wealth of myth and folklore associated with cassowaries is a most curious conviction fostered by such tribes as the Huri and the Wola from Papua New Guinea's remote Southern Highlands Province. According to their lore, a female Bennett's cassowary maintained in captivity is able to reproduce even if she is not provided with a male partner. All that she has to do is locate a specific type of tree and thrust her breast against its trunk, again and again, in an ever-intensifying frenzy, until at last she collapses onto the floor in a state of complete exhaustion, suffering from internal bleeding that festers and clots to yield yellow pus. This in turn proliferates, producing yolk-containing eggs that the female lays, and which are incubated and hatch as normal.

Although a highly bizarre tale, it is worth recalling that cases of parthenogenesis (virgin birth) are fully confirmed from a few species of bird, notably the common turkey; the offspring are genetically identical to their mother. Perhaps, therefore, this odd snippet of native folklore should be investigated—just in case (once such evident elements of fantasy as the pus-engendered yolk are stripped away) there is a foundation in fact for it, still awaiting scientific disclosure.

An even more imaginative Wola belief regarding Bennett's cassowary concerns its migratory habits. As revealed by Paul Sillitoe during a filming expedition to Wola territory in 1978 (*Geographical Journal*, May 1981), these birds only visit this area when the fruits upon which they feed are in season here. At the season's end they travel further afield again, but the Wola are convinced that they have gone to live in the sky with a thunder goddess (though they neglect to reveal how these flightless birds become airborne!).

Irrespective of these charming tales, it is true that for flightless birds the cassowaries do exhibit an extraordinarily dispersed, far-flung distribution—occurring on a surprising number of different islands. Admittedly, many of these islands were once joined to one another in the not-too-distant geological past, but some ornithologists remain doubtful that the cassowaries' range is natural—suggesting instead that they may have been introduced onto certain of their insular territories by humans. For example, Drs. A.L. Rand and Thomas Gilliard proposed in their *Handbook of New Guinea Birds* (1967) that *C. casuarius* may well have been brought by man to Seram. In view of the New Guinea tribes' very extensive trade in cassowaries—not only transporting them across land but also exporting them far and wide in boats, a tradition known to have been occurring for at least 500 years, such a possibility is by no means implausible. It was raised in 1975 by Dr. C.M.N. White too, within the British Ornithologists Club's bulletin, and he offered a corresponding explanation in the same publication the following year for the presence on New Britain of Bennett's cassowary.

Probably the most unexpected variation on the theme of displaced cassowaries, however, is a case aired by Drs. G.H. Ralph von Koenigswald and Joachim Steinbacher (*Natur und Museum*, 1986), who reported the presence of a bas-relief depiction of a readily identifiable cassowary upon a temple near Wadjak in eastern Java. As there is no evidence to imply that Java ever harbored a native form of cassowary, this depiction lends itself to a variety of different cultural interpretations. For example, it suggests that the centuries-old tradition of cassowary trade and export from New Guinea even extended as far afield as Java—the ownership of one such bird by the ruler of Lydajo, Java, in 1597 bolsters this hypothesis. Alternatively, the depiction might simply have been based upon *descriptions* of cassowaries, recounted to the Javan natives by visiting New Guinea traders. There is even the chance that the Javan tribe responsible for this painting was descended from one that had migrated to southeast Asia from New Guinea, and the picture was inspired by orally preserved traditions among this translocated people of birds known to their ancestors in New Guinea.

Incidentally, comparable hypotheses to those outlined above for the "Javan cassowary," but this time featuring travelers or ancestors from New Zealand rather than from New Guinea, could also explain the moa-like bird depicted upon the 800-year-old principal temple of Angkor in Cambodia—another iconographical oddity unfurled by von Koenigswald and Steinbacher in their paper.

DREAMS OF A FEATHERED *GERONTICUS*

The famous Greek legend featuring Heracles and the dreaded Stymphalian birds. A non-existent forest raven native to the lofty peaks of the Swiss Alps. The epic biblical story of Noah and the Great Flood. What conceivable connection could exist between such ostensibly disparate subjects as these? The answer is a genuine *rara avis* called *Geronticus*—a baffling bird encompassed by all manner of myths and mysteries, drawn together from the dreams of the past.

One of the twelve great labors imposed upon the mighty Greek hero Heracles by the cowardly King Eurystheus of Argolis was to vanquish the Stymphalides—a flock of brass-winged, crane-sized, ibis-like birds with crests and a craving for human flesh, which frequented the Stymphalian marshes in Arcadia. With the aid of a loudly resonating pair of brass rattles forged by the fire god Hephaestus and loaned to him by the war goddess Athena, however, Heracles was able to scare the birds out of their marshy domain and into flight. Once they were airborne, wheeling over-head in a frenzy of fright engendered by the ear-splitting cacophony of the rattles, he then proceeded to fire volley after volley of arrows into their soft, unprotected bellies, killing some and sending the remainder flying far away toward the Black Sea, never to return to Arcadia.

These fearful bird-monsters were originally thought to constitute an imaginative personification of marsh fever, but in 1987 Swiss ornithologist Michael Desfayes proposed that they may actually have been based upon a real (albeit smaller, harmless) bird.

Also called the hermit ibis, the waldrapp *Geronticus eremita* is a very rare relative of the more familiar sacred ibis of Egypt and South America's scarlet ibis, and sports a crest, a long red beak, and bronze-colored wings. It is nowadays restricted to a scattering of sites in Morocco and Algeria, and one notable locality in Turkey—its only toehold in Europe. However, it was once much more wide-spread, known from as far west as Germany, Austria, and

Switzerland (see below). Moreover, after comparing its morphology with classic descriptions of the Stymphalides, Desfayes opined that they were one and the same species, and that Greece should thus be added to the waldrapp's former distribution range.

Another longstanding mystery bird from Europe was the Swiss forest raven described and portrayed by Zurich scholar Conrad Gesner in his *Historiae Animalium,* published in 1555. Later writers dismissed this bird as imaginary, because it seemed to resemble a crested ibis rather than a raven, and there was no ibis-like bird known from Switzerland. Not until 1941 was the mystery finally resolved, when geologically Recent remains of waldrapps were excavated from the Glarus Alps and sites near Solothurn.

This distinctive bird corresponds perfectly with Gesner's description and depiction of his perplexing "forest raven," and it is now known that way back in 1504 the waldrapp was formally designated a protected species in the Alps by Archbishop Leonard of Salzburg. Tragically, however, his decree was largely ignored, and the

GESNER'S "FOREST RAVEN"

waldrapp's nestlings were hunted mercilessly as prized culinary delicacies. Less than a century later it had become extinct here, swiftly fading from memory and ultimately transforming into an apparently mythical, non-existent bird known only from a medieval bestiary.

It is likely that the waldrapp would have become extinct in Turkey too, but here it has been saved not only by modern conservationist zeal but also by traditional religious belief. According to ancient Turkish lore, the waldrapp was one of the birds released by Noah after the Great Flood, and symbolizes fertility—giving the people of Birecik, site of this species' only notable Turkish colony, an added incentive for perpetuating its tenuous survival here.

HUGE HORNBILLS AND TREMENDOUS TAILS

The Torres Strait separates New Guinea from the northernmost tip of Queensland, and contains many tiny, virtually unexplored islets. According to the 19th century naturalist Luigi d'Albertis, among others, one of these houses a huge species of hornbill, known as the *kusa kap*. The largest *known* species include the familiar great Indian hornbill *Buceros bicornis* and the rhinoceros hornbill *B. rhinoceros* (from Malaysia, the Sundas, and Borneo), both of which can attain a total length of four feet, but these pale into insignificance alongside the *kusa kap*—whose alleged 22-foot wingspan, not to mention its inclination for carrying dugongs aloft in its claws, more closely recall the legendary elephant-transporting roc! Moreover, the noise of its wings flapping in flight is said to resemble the roar of a steam engine!

Although I think it exceedingly unlikely that the *kusa kap*, at least in the form of the above description, will ever prove to be anything more than an inhabitant of the imagination, it is worth noting that the rhinoceros hornbill *is* known to produce a sound with its wings that has been likened by listeners to the noise of a chugging steam train. Perhaps, then, the *kusa kap* is an undescribed relative of this species, still awaiting official discovery on one of the Torres Strait's islets, and whose dimensions have been greatly exaggerat-

RHINOCEROS HORNBILL

ed in local traditions. It would not be the first time that an impos-
sibly flamboyant animal of native legend has been shown to conceal
at its core a much more sober but indisputably real species in need
of formal scientific recognition.

The same may be true of the superficially improbable bird
spied by a Mr. Sporing in the 18th century on New Zealand's Jubolai
Island. According to a record of this sighting in the journal of Joseph
Banks (Captain Cook's Botany Bay naturalist) for October 1769:

While Mr. Sporing was drawing on the island he saw a most strange bird fly over his head. He described it as being about as large as a kite, and brown like one; his tail, however, was of so enormous a length that he at first took it for a flock of small birds flying after him: he who is a grave thinking man, and is not at all given to telling wonderful stories, says he judged it to be yards in length.

Were it not for the fact that there are no such species native to New Zealand, it would be tempting to identify this form as one of those excessively long-tailed birds of paradise known as astrapias, indigenous to New Guinea. The most spectacular of these, the ribbon-tailed bird of paradise *Astrapia meyeri*, was not even discovered until 1938; the male has only a tiny body, but sports an extraordinary tail composed of two enormous white feathers—each of which is at least a yard long.

MAORI MAN-EATERS AND CONGOLESE MONKEY-SNATCHERS

New Zealand was once home to two very impressive eagles reminiscent of the formidable South American harpy eagle *Harpia harpyja*, but with longer limbs. Known respectively as *Harpagornis moorei* and *H. assimilis,* and believed from the evidence of various incomplete remains to have been stronger and more heavily built than any of today's eagles, with a wingspan of around seven feet, these mighty raptors were once thought to have died out as long ago as the 12th century. However, Maori legends tell of two gigantic, terrifying birds referred to as the *hokioi* and the *poua-kai* (this latter was supposedly a man-eater, according to South Island's Waitaha tribe), whose descriptions tally closely with reconstructions of the *Harpagornis* species from fossils. In fact, many ornithologists now consider it possible that one or both of these eagles persisted until as recently as the 17th century—and that if it had not been for the extermination by humans of their likely prey, the smaller species of moa, they might have survived even longer.

There is no doubt that they must have been a breathtaking sight in life, but a quite recent discovery showed that they were even more spectacular than hitherto suspected. At the beginning of 1990, it was announced that the first *complete* skeleton of a *Harpagornis* eagle had been found, concealed deep inside a cave of Mount Owen, near Nelson, South Island (*Daily Telegraph,* January 17, 1990). From this extremely important find, an adult *H. moorei,* scientists were able to deduce that in life its species would have weighed about 35 pounds, and had a wingspan of up to 10 feet. Little wonder, then, that *Harpagornis* has been preserved in the Maoris' memories—the sight of a predatory bird soaring overhead on wings spanning 10 feet across is not one to be readily forgotten!

In another part of the world, moreover, such a sight can still be made today—at least according to the natives of the Peoples Republic of the Congo, who firmly believe in the existence of an enormous type of monkey-devouring eagle called the *ngoima.* During his searches in the 1980s for this country's most famous mystery beast, the swamp-dwelling dinosaur-like *mokele-mbembe,* cryptozoologist Dr. Roy P. Mackal also received details of the great *ngoima* bird, from one of its most eminent eyewitnesses—André Mouelle, the country's political commissar. Mouelle stated that it had a huge wingspan of nine to 13 feet, preyed upon monkeys and goats, dwelt within the dense forests, and was armed with talons as large as a man's hands, legs as big as a man's forearms, and a downward-curving hooked beak. Its plumage was blackish-brown above, with somewhat paler underparts.

As Mackal discusses in his book *A Living Dinosaur?* (1987), the *ngoima* corresponds quite closely both in appearance and in prey preference to the martial eagle *Polemaetus bellicosus* (if we assume that the upper limit of the *ngoima's* wingspan as given by Mouelle is an overestimate); but whereas the *ngoima* is reputedly a forest dweller, the martial eagle is a bird of open country. In contrast, there is another, albeit slightly smaller species that shares not only the *ngoima's* morphology and prey preference but also its forest-dwelling habitat. This is the crowned eagle *Stephanoaetus coronatus*, the most powerful (yet also the least-observed) eagle in Africa. Indeed, it corresponds so closely with the *ngoima* on every count other than the latter bird's immense wingspan that Mackal considers the search for the *ngoima's* identity to be at an end, with the dramatic dimensions of its wings merely the product of human exaggeration (when confronted unexpectedly by the daunting presence of this rarely spied species), rather than any legitimate manifestation of nature.

Nevertheless, as revealed in this chapter, there are plenty of other ornithological oddities still in need of satisfactory explanations—enough, in fact, for even the most adventurous bird-watcher to get into a flutter about!

CHAPTER 10

Slithery Surprises

Forth from the dark recesses of the cave
The serpent came; the Hoamen at the sight
Shouted; and they who held the priest, appall'd,
Relaxed their hold. On came the mighty snake,
And twined in many a wreath around Neolin,
Darting aright, aleft, his sinuous neck,
With searching eye and lifted jaw, and tongue
Quivering; and hiss as of a heavy shower
Upon the summer woods. The Britons stood
Astounded at the powerful reptile's bulk,
And that strange sight. His girth was as of man,
But easily could he have overtopp'd
Goliath's helmed head; or that huge king
Of Basan, hugest of the Anakim.
What then was human strength if once involv'd
Within those dreadful coils! The multitude
Fell prone and worshipp'd.

ROBERT SOUTHEY—*MADOC. THE CURSE OF KEHAMA*

ALL THAT GLITTERS IS NOT GOLD. AND ALL THAT
slithers is not snake—or is it? Peruse the following pentet of limbless—or near-limbless—cryptids, and judge for yourself!

THE JUMPING SNAKE OF SARAJEVO

Sometimes, the briefest words attract the greatest interest. So it was when, in the closing section of my book *In Search of Prehistoric Survivors* (1995), I included the following line: "I have on file reports of a mysterious jumping snake reported from the environs of Sarajevo." Since then, I have received numerous communications from readers requesting further details—more, in fact, than for almost any other mystery beast noted by me anywhere. So here, at last, is what I know about the Sarajevo jumping snake.

In reality, this mystifying serpent has been reported widely across the old, pre-fragmented Yugoslavia, but I first learned about it from a chapter on alleged jumping snakes in Dr. Maurice Burton's book *More Animal Legends* (1959). Burton quoted a lengthy letter from Mr. M.F. Kerchelich of Sarajevo, which included the following important excerpts:

> We have in Yugoslavia a dreaded specimen
> [species] of jumping snakes called locally
> 'poskok', meaning 'the jumper'…found mainly in
> Dalmatia along the east coast of the Adriatic and
> the mountainous regions of Hercegovina [sic] and
> Montenegro. In fact, there exists an island near
> Dubrovnik called 'Vipera', which is a well-known
> breeding place of 'the jumper'. The average size
> of this snake is between 60 and 100 cm, though
> considerably larger specimens were seen. The
> colour seems to vary, according to environment,
> from granite grey to dark reddish brown. The
> snake is dreaded for its poison and aggressive-
> ness when disturbed, though it will usually hide
> on approach of man…I can vividly remember
> three cases of meeting the 'poskok', the first was
> in Montenegro, when on a trip I stopped the car
> to stretch my legs. It was lying in the middle of

the road, some 100 metres ahead of the car, sleeping in the sun. After watching it for some time it must have felt my presence, curled up and jumped into a dry thorn bush at the kerb of the road and disappeared. The jump was at least 150 cm long and some 80-90 cm high. On the second occasion I was driving a jeep in southern Dalmatia and, coming round the corner, I could just see a snake about to cross the road. The car must have frightened it and quite suddenly it jumped at the front mudguard and was killed by the rear wheel. The third time I met a 'poskok' was when I was fishing a small stream in Croatia and leaning against a dry tree trunk. Suddenly I noticed a hissing sound above me, and, having been warned of snakes, scrutinised my surroundings more carefully. It took me quite a time to discover that one of the dry branches was not a branch at all but a 'poskok' watching me intently…All three specimens I have seen were not more than about 3 to 4 cm in diameter at the thickest point.

Burton gave reports of jumping snakes from elsewhere, too, including his own sighting of a grass snake leaping across a ditch that was one foot deep and two feet wide, and his experience of another grass snake apparently leaping with lightning speed over the wall of a vivarium, by stiffening its body in a vertically erect position. Burton concluded: "the ability [of snakes] to progress in vertical jumps becomes at least feasible. All that is needed is the stimulus."

Thus, Burton did not consider the *poskok's* disputed jumping ability to be zoologically impossible. However, he offered no opinion whatsoever regarding its taxonomic identity.

I later mentioned the *poskok* to London University zoologist

Professor John L. Cloudsley-Thompson, an expert in reptilian behavior. In a paper titled "Reptiles That Jump and Fly," published by the *British Herpetological Society Bulletin* in 1996, he commented: "cryptozoologists have reported a mysterious jumping snake from the environs of Sarajevo. Could this be *Coluber viridiflavus*, the western whip snake?"

Unfortunately, this species' distribution only extends into the northwestern section of the former Yugoslavia. However, three of its close relatives—the Balkan whip snake *C. gemonensis* (which greatly resembles it), the large whip snake *C. jugularis*, and Dahl's whip snake *C. najadum*—all occur much more widely in Yugoslavia. Exhibiting quite a range of colors, including those reported for the *poskok*, these species are of similar size to it too, and the large whip snake can also attain the larger sizes sometimes claimed for it. Moreover, all are very slender, fast-moving, aggressive when captured, and given to biting hard. In stark contrast to the *poskok*, conversely, they are not venomous.

Coupling its venom with the name of the Dubrovnik island—Vipera—where the *poskok* is said to breed, I favored one of Europe's several species of viper as a more likely candidate (always assuming, of course, that it does not comprise a hitherto undescribed species). Three vipers are known to occur widely in the former Yugoslavia—Orsini's viper *Vipera ursinii*, the common viper (adder) *V. berus*, and the nose-horned viper *V. ammodytes*. Moreover, one of these—the nose-horned viper—is highly venomous.

In August 1998, I was contacted by Marko Puljic, of Yugoslavian descent, who had read my brief snippet in *Prehistoric Survivors*, and had afterwards communicated with Croatian herpetologist Dr. Biljana Janev. In her reply to Marko, Dr. Janev stated that the *poskok* was indeed the nose-horned viper, continuing: "Its name [poskok—'jumper'] comes from the belief that it can jump [far], which is not accurate. Rather, it is able to jump like many other snakes (maybe 0.5 metre, and that is the 'lunging' during an attack). Maybe the reason is that the snake, particularly in autumn, climbs on shrubs and low trees where it can hunt young birds. Believably [Presumably], if it attacks, it looks like it has jumped quite a way."

This explanation of the *poskok's* jumping ability is the traditional solution offered by skeptics of reports concerning jumping snakes, but as witnessed by Burton with grass snakes, it cannot explain all such cases. Moreover, it seems strange that Kerchelich's detailed *poskok* description did not mention the striking dorsal zigzag markings or nose-horn sported by *V. ammodytes*. And how do we explain the testimony of a relief worker in Metkovic (southern Croatia, on the Bosnian border), speaking to Marko in October 1996, who stated that he always warned the British SFOR (NATO Peacekeeper) soldiers to take care where they trod as a snake exists here that can jump six feet into the air? Folklore, fancy...or fact?

TRAILING THE *TATZELWORM*

Thoughts of mystery animals normally conjure up images of remote tropical localities far from civilization, rarely visited by Westerners. In reality, however, one of the most tenaciously elusive mystery beasts has been reported for centuries from the very heart of Europe — the Alps mountain range, extending through Switzerland, Austria, Italy, and Bavaria. It remains unknown to science, but is well known locally as the *tatzelworm* ("clawed worm"), *stollenworm* ("hole-dwelling worm"), or *springworm* ("jumping worm").

Many eyewitness accounts of *tatzelworms* have been documented, some containing conflicting morphological details. Generally, however, this enigmatic beast can be described as follows:

Measuring two to four feet long, with light-colored skin usually appearing smooth or surfaced with tiny scales, the *tatzelworm* is very elongate in shape, with a short head often said to be cat-like in form, little or no neck, a long worm-like body, and a blunt tail. Some observers claim that it has a pair of short, clawed limbs at the front of its body; others state that it also has a second, hind pair of limbs; and a few allege that it has no limbs at all. It reputedly lives mainly in holes or burrows, but is sometimes encountered basking in alpine meadows on sunny days. If approached, it will usually vanish swiftly into its hole, but there are reports of decidedly belligerent *tatzelworms* that have leaped toward their startled eyewitnesses, usually causing them to flee rather than continue their observations!

Although the existence of such a sizeable creature still undiscovered by science in the Alps may seem unlikely to outsiders, the *tatzelworm's* reality to these mountains' inhabitants is such that in the 1800s it was even featured in three major alpine guides. According to one of them, Swiss naturalist Friedrich von Tschudi's *Das Thierleben der Alpenwelt* (1861):

> In the Bernese Oberland and the Jura the belief is
> widespread that there exists a sort of 'cave-worm'
> which is thick, 30 to 90 cm long, and has two
> short legs; it appears at the approach of storms
> after a long dry spell.

A *tatzelworm* drawing matching the above description had appeared 20 years earlier in a Swiss almanac, *Alpenrosen*. However, perhaps the most famous *tatzelworm* illustration is the sketch of a "scaly log" with a toothy grin, plus front and rear limbs, featuring in a Bavarian handbook, *Neues Taschenbuch für Natur-, Forst- und Jagdfreunde auf das Jahr 1836,* and based upon the description given by someone claiming to have shot one of these creatures.

Over the years, some fascinating *tatzelworm* sightings have been recorded. One of the most dramatic, however, must surely have been the decidedly too-close encounter of the scary kind experienced one day in summer 1921 by a poacher and a herdsman while hunting on Austria's Hochfilzenalm Mountain, near Rauris. Suddenly, they became aware of a very strange animal, lying on a rock close by, watching them. Measuring two to three feet long, it was grey in color, as thick as a human arm, with a feline, fist-sized head, and a thick tail. Alarmed, the poacher raised his rifle and shot at this bizarre beast—but as he did so, the creature, living up to its alternative name of *"springworm,"* abruptly jumped up through the air toward them, revealing two small front limbs but no rear limbs. Not surprisingly, the men raced away.

MODEL OF *TATZELWORM*
(Ivan Mackerle)

A pallid smooth-skinned specimen measuring approximately 18 inches long with two stub-like front limbs and noticeably large eyes was encountered in April 1929 by a teacher seeking the entrance to a cave on Austria's Mount Landsberg. The creature, which may have been a juvenile specimen because of its small size, was lying in wet, moldy leaves, but swiftly vanished into a hole close by when the teacher tried to capture it.

Intriguingly, together with a number of snakes, an apparent *tatzelworm* was discovered in 1933 concealed inside a hollow space behind a stone wall being removed by some workmen at Spittal an der Drau in Austria. According to their description, it was three feet long, dirty white in color with a yellowish tinge, and cylindrical in shape, with a cat-like head, big eyes, and two short front legs. The workmen gingerly maneuvered this peculiar animal onto a shovel and, along with its serpentine companions, tossed it into the nearby Lieser river—across which it rapidly swam, until it disappeared from sight on the far-side bank.

Another noteworthy sighting took place near St. Pankraz, Austria, in 1922, when a 12-year-old girl playing in a wooded area ran to see why her sister, playing close by, had suddenly begun to cry. When she reached her, the girl was terrified to spy, no more than four yards away, a grey-colored creature with transverse grooves across its body, crawling between some stones. Measuring at least one foot long, it resembled a giant worm, but sported a pair of paws behind its head. Too shocked to move, the two sisters gazed at it in fascinated horror for a time before summoning up enough courage to run away.

One of the earliest but most memorable accounts seemingly appertaining to a *tatzelworm* is that of Jean Tinner, who was walking with his father on Switzerland's Frumsemberg Mountain one day in or around 1711 when they encountered an extraordinary snake-like creature coiled up on the ground. Measuring seven feet long, it was apparently limbless, but was readily distinguished from normal snakes by its remarkably cat-like head. Taking no chances with such an uncanny creature, Tinner shot it, but its body was not preserved.

Ironically, several *tatzelworm* specimens have apparently been obtained, yet all have been discarded. One incomplete four-foot-long skeleton was allegedly found in Austria's Mur Valley during 1924, but was dismissed as a partial roe deer skeleton by a veterinary student. However, this identification was disputed by its finder, and two years later, in the exact area where the skeleton had been found, a young shepherd encountered a large reptilian mystery beast which so frightened him that he refused to go back there that year. A smaller, three-foot-long specimen has also been spied alive in this vicinity.

One winter sometime prior to 1910, a farmer in Italy's South Tyrol mountain range unexpectedly discovered a torpid *tatzelworm*-like beast asleep (hibernating?) in some hay on his farm. Inevitably, he promptly dispatched it, but did not retain its body, noting only that after he had killed it a green liquid seeped out of its mouth.

And in 1828, a peasant from Switzerland's Solothurn canton came upon a dead *tatzelworm* in a dried-up marsh, and put it aside, in order to present it for scientific examination. Tragically, however, before he could do so, half of it was devoured by some crows, and when its skeleton was eventually sent to Heidelberg it was duly lost!

A number of candidates have been offered. Many cryptozo-oologists favor some form of elongate lizard, such as a skink, some species of which are worm-like with very reduced limbs. Moreover, certain skinks only have a single pair. Another popular identity is an unknown species of legless lizard allied to the slow worm *Anguis fragilis* and Europe's larger, inappropriately dubbed glass snake *Ophisaurus apodus*. There have even been attempts to link it to the exclusively New World Gila monster *Heloderma suspectum* and its close relative the Mexican beaded lizard *H. horridum*—the world's only known venomous lizards.

There are also three species of extraordinary reptiles called ajolotes (see also Chapter 2), which resemble large earthworms, except for their small pair of front paws—a description corresponding very closely indeed with the *tatzelworm*. Unfortunately, ajolotes are known only from Mexico.

Alternatively, some researchers prefer an amphibian identity, comparing the *tatzelworm* with certain salamanders, such as the American sirens, which only possess front limbs—or the amphiumas, also lacking hind limbs and only possessing minute front limbs.

My own feeling is that a reptilian identity is more probable, as some *tatzelworms* have been recorded far from water, whereas amphibians typically prefer damp surroundings.

However, it is possible that speculation concerning the *tatzelworm's* identity is already academic, as sightings seem to have become far less frequent in recent years, indicating that this mysterious entity, even if it does exist, may now be an endangered species.

In June 1997, Czech cryptozoologist Ivan Mackerle led a week-long expedition to the Austrian Alps in search of the *tatzelworm*, visiting Hochfilzenalm and the Mur Valley, but was saddened to learn that only the more elderly interviewees could recall it and describe it, again suggesting either that its tradition, or the creature itself, is dying out.

How tragic it would be if a mystery beast widely reported for centuries in Europe should have become extinct before Western science had even recognized its existence.

GOING LOCO FOR A *TZUCHINOKO*

Scotland has Nessie, Canada has Ogopogo, and Mongolia has its death worm. But the mystery beast that keeps Japanese cryptozoologists on the lookout is the *tzuchinoko*. Known here for centuries, this bizarre snake is fairly short but very thick-bodied, and reputedly resembles an animated bar of Toblerone chocolate—inasmuch as its prominent dorsal ridge and flattened undersurface give it a triangular cross-sectional shape when viewed head-on.

It also boasts a pair of squat horn-like projections above its small eyes, well-defined facial pits, and a clearly delineated neck. Cryptozoological researchers Michel Dethier and Ayako Dethier-Sakamoto have speculated that if such a snake does exists, it may be a mutant form of the pit viper *Agkistrodon halys*, or possibly even a totally new, undescribed species.

TWO VIEWS OF THE *TZUCHINOKO*
(International Society of Cryptozoology)

During recent years, certain Japanese towns have purposefully flaunted the *tzuchinoko's* curio appeal as a tourist lure—by offering substantial monetary rewards to anyone who can capture a specimen. In 1992, the mountain ski village of Chigusa offered 200 million yen (then equal to £490,000) for a living *tzuchinoko,* and 100 million yen for a dead one. But nothing has been forthcoming—until August 2000, that is, when the Japanese media reported that a *tzuchinoko* body had been procured in Yoshii, a town in central Okayama Prefecture. As so often happens in cryptozoology, however, the truth proved to be rather different.

It all began on May 21, 2000, when, according to a *Mainichi Shimbun* [Daily News] account, a farmer cutting grass claimed to have seen "a snake-like creature with a face resembling cartoon cat Doraemon[!]" slither across a field. The farmer slashed at it with his weed whipper, but the creature escaped into a nearby stream. It had evidently been fatally wounded, however, because on May 25, its dead body was found lying beside the stream by 72-year-old Hideko Takahashi, who buried it. Describing its appearance to the newspaper, she recalled that "it had a cute, round face and was clearly not a snake. I've seen something like this around here before, and it makes a chirping sound."

When the Yoshii Municipal Government heard about this, it dispatched some officials who located the creature's burial site and dug up its remains, forwarding them to Prof. Kuniyasu Satoh at Kawasaki University of Medical Welfare to examine and identify. Meanwhile, *tzuchinoko* fever hit Yoshii in a big way. Its mayor, Ryuichi Arashima, announced that he had no doubts of this mystery species' existence here, claiming that many sightings had been reported locally in the past. Accordingly, he offered a reward of 20 million yen to anyone who caught a live specimen, and even promised to increase the prize by a million yen a year until one was captured. Predictably, Yoshii was soon awash in bounty-hunting tourists, anxious to bag one of these elusive beasts. But even if they didn't (which they didn't), they could console themselves with a bottle of *tzuchinoko* wine, a *tzuchinoko* rice cake, or even a *tzuchinoko* pastry, courtesy of some decidedly enterprising local enterprises.

Even the municipal government lost no time in entering the fray—by formally launching a project team. Resplendently outfitted in special *"tzuchinoko* hunter" caps and uniforms, the team duly interviewed any elderly Yoshii citizens claiming to have seen *tzuchinokos* in their youth, and patrolled the town hoping for fresh sightings.

One of the most memorable testimonies to emerge from this investigation was that of 82-year-old Yoshii resident Mitsuko Arima, who allegedly sighted a *tzuchinoko* as recently as the morning of June 15, swimming along a river:

> I was surprised. I just pointed at it and asked
> 'Who are you? Who are you?'. It didn't answer
> me, but just stared. It had a round face and
> didn't take its eyes off me. I can still see the eyes
> now. They were big and round and it looked like
> they were floating on the water. I've lived over
> 80 years, but I'd never seen anything like that in
> my life.

By September, however, the great *tzuchinoko* hunt was quietly petering out. Moreover, the final news reports stated that the dead specimen had merely been identified as "a kind of snake"—a typically vague, inconclusive end to another crypto-case. Yet as far as I could see, this entire episode had had little if anything to do with the "real" *tzuchinoko* anyway. After all, the recent eyewitness descriptions given here bear scant resemblance to the classic *tzuchinoko* description cited at the beginning of this article. They report a creature with a round face instead of a triangular one, large eyes instead of small ones, and include no mention whatsoever of any of the *tzuchinoko's* most characteristic features—horns, triangular cross-sectional shape, distinct neck, flattened underside, and short thick body.

If only a more detailed account of the dead specimen examined by Prof. Satoh could be obtained, to clarify this much-muddled situation. Happily, such an account was indeed obtained, by Gavin Joth, who succeeded in contacting Satoh, and kindly presented his informative reply online to fellow members of the cz@egroups.com discussion group.

Satoh revealed that he had been able to reconstruct the dead snake's bone structure from its remains, and he estimated its total length at 120 cm (four feet). Its largest ventral scales were 49.5 mm (roughly two inches) across, and the morphology of its dorsal scales corresponded with those of Japan's tiger water snake *Rhabdophis tigrinus*. Also consistent with this species was a small fang found on the dead snake's maxillae. Thus Satoh concluded that this was indeed the identity of the latter specimen (though conceding that it was a notably large one).

As this species is very visibly dissimilar from the classic *tzuchinoko* description, which is much closer to that of pit vipers, the good citizens of Yoshii were presumably led astray in their assumptions that they had sighted *tzuchinokos* in 2000—seemingly by a snake in the grass that proved to be a serpent of the water, and not a tobleronic *tzuchinoko* after all.

NEVER BOTHER A *BEITHIR!*

When is an eel not just an eel? When it is a mysterious Scottish snake, or the Loch Ness monster, or a bottled sea serpent, or a giant blue worm from India, or...

Monstrous eels far greater in size than any officially recorded by science have been offered at one time or another as a favored identity for a wide range of cryptozoological creatures still awaiting formal discovery. Take, for instance, the *beithir*.

One of Britain's lesser-known mystery beasts, the *beithir* is said to resemble a giant water snake, and reputedly inhabits the dark caves and waters of the Scottish Highlands' more remote, secluded areas. A correspondent in the English magazine *Athene* recalled meeting a fisherman near Inverness in 1975 who claimed that he and four others had once seen a *beithir* lying coiled in shallow water close to the edge of a deep gorge upstream of the Falls of Kilmorack. When the creature spied them, however, it began to thrash violently, and then swam up the gorge near Beaufort Castle, eventually vanishing from sight. They estimated it to have been "a good three paces or more" long, i.e. nine to ten feet.

Equally intriguing is the testimony of a Strathmore game-keeper who alleged that his wife's parents had spied *beithirs* moving *on land* during the 1930s at Loch a' Mhuillidh, near Glen Strathfarrar and the mountain of Squrr na Lapaich. Could these elongate enigmas be over-sized (or over-estimated) specimens of the grass snake *Natrix natrix*, a species equally at home on land and in the water? Or is it possible that they are unusually large eels, whose abilities to migrate considerable distances overland are well known?

Any contemplation of mysterious water beasts in Scotland leads inevitably to Nessie—the famously elusive monster of Loch Ness. And this too has been provisionally identified by some zoological authorities, notably the late Dr. Maurice Burton, as a giant eel, possibly up to 30 feet in total length. Normally, of course, the common European eel *Anguilla anguilla* does not exceed five to six feet, and the conger eel *Conger conger* seldom exceeds nine to ten feet. Nevertheless, ichthyological researchers have revealed that growth in eels is more rapid in confined bodies of water (such as a loch), in water that is not subjected to seasonal temperature changes (a condition met with in the deeper portions of a deep lake, like Loch Ness), and is not uniform (some specimens grow much faster than others belonging to the same species).

Collectively, therefore, these factors support the possibility that abnormally large eels do indeed exist in Loch Ness. Also of significance is the fact that eels will sometimes swim on their side at or near the water surface, yielding the familiar humped profile described by Nessie eyewitnesses, and a 20-foot or 30-foot eel could certainly produce the sizeable wakes and other water disturbances often reported for this most celebrated of all aquatic monsters.

Even so, independent sightings of the beast seen in its entirely, i.e. on land, have repeatedly featured a creature possessing distinctive flipper-shaped limbs, a well-delineated neck, a burly body, and a long tail. This description is far removed indeed from that of any eel, but is strikingly reminiscent of a supposedly long-extinct type of aquatic reptile known as a plesiosaur—which remains the most popular identity among cryptozoologists for the Loch Ness monster. Yet even if Nessie is a plesiosaur, there is no reason why Loch Ness could not *also* harbor some extra-large eels. After all, any loch that can boast a volume of roughly *263 billion* cubic feet must surely have room enough for more than one monster!

Immense landlocked eels supposedly inhabit a number of deep pools in the Mascarene island of Reunion, near Mauritius. In a letter to *The Field* (February 10, 1934), Courtenay Bennett recalled seeing in the 1890s a dead specimen that had been caught in one such pool, the Mare à Poule d'Eaux, and from which "steaks as thick as a man's thigh were cut."

Some cryptozoologists believe that certain extremely elongate sea monsters sporadically reported over the years may comprise a hitherto-undiscovered species of gigantic eel. Indeed, it has even been given a name, the super-eel—and for many years, scientists were convinced that a young super-eel had actually been captured and preserved.

The curious case of the bottled sea serpent began on January 31, 1930, when the Danish research vessel *Dana* caught a truly extraordinary juvenile eel (leptocephalus) at a depth of about 900 feet, south of the Cape of Good Hope. What made this specimen so astonishing was its size. In contrast to the common eel's leptocephalus, which is no more than three inches long (even the conger eel's is only four inches long), this colossal leptocephalus measured six feet, one and a half inches!

Bearing in mind that during its metamorphosis from leptocephalus to adult, the common eel becomes 18 times longer (and the conger can become as much as 30 times longer), bemused ichthyologists estimated that the still-unknown adult version of the mysterious species represented by the *Dana* leptocephalus could be anything between 108 and 180 feet long! A veritable super-eel in every sense. Accordingly, this highly significant leptocephalus was duly preserved in alcohol, and is now housed within Copenhagen University's Zoological Museum. Meanwhile, the zoological world anxiously awaited the capture of an adult super-eel.

Alas, it was not to be. In 1970, Miami University ichthyologist Dr. David G. Smith conclusively identified the giant leptocephalus's species as a spiny eel or notacanthid. Despite their name, spiny eels are not true eels; moreover, unlike true eels, which undergo most of their growth *during* metamorphosis, spiny eels undergo most of theirs *prior to* metamorphosis. Consequently, had it lived and transformed into an adult, the six-foot-long *Dana* leptocephalus would not have greatly increased in size—thus sweeping aside earlier speculations that it could have become a 100-foot-plus monster adult eel.

Having said that, it is by no means impossible that a gigantic species of true eel does indeed await formal scientific discovery; the ocean depths are far too immense for even the most conservative opinion to dismiss such a prospect out of hand. At present, however, there is only tantalizing anecdotal evidence for the existence of such creatures.

And what can be said about the giant worm-like eels (or eel-like worms?) with vivid blue bodies that reputedly lurk amid the dank riverbed ooze of the Ganges? This, at least, is what Aelian, Ctesias, Solinus, and a number of other celebrated scholars from the ancient world once claimed. According to Solinus, these amazing creatures were 30 feet long, but their dimensions grew ever greater with repeated retellings by subsequent writers until they eventually attained sufficient stature to emerge from their muddy hideaways at night and devour unwary camels and cattle! Not surprisingly, this

incredible species of eel has never been brought to scientific attention—a classic example, presumably, of the one that got away!

THE *NAGA*—MONSTROUS SNAKE
OF THE MEKONG

In Hindu and Buddhist mythology, a *naga* is an ancient serpentine
deity, sometimes portrayed as an entity combining the body of a
huge snake with the head (and sometimes the torso too) of a human,
and an erectile scaly crest resembling the hood of a great cobra. In
cryptozoology, however, *"naga"* has a rather different meaning as
the name given to an alleged giant water snake that inhabits the
Mekong River constituting the northeastern border of Thailand with
Laos. In October 2000, British cryptozoologist Richard Freeman visited Thailand with a film crew to seek, and, if possible, spy or even
film this mysterious beast.

Sadly, neither a film nor any firsthand sightings materialized,
but Richard did succeed in collecting some very interesting eyewitness accounts and information about the cryptozoological *naga*
that had not previously been made available to Western researchers.

One such eyewitness was none other than the chief of police
in Phon Pisai village, Officer Suphat. In 1997, he and a group of thir-
ty people had been walking along some cliffs overlooking the
Mekong when, looking down, they saw what at first seemed to be
flotsam, floating on the river. Peering closer, however, they were
shocked to discover that this "flotsam" was actually a truly
immense black snake, swimming against the river's current via a
series of powerful horizontal body undulations. Officer Suphat estimated that it measured no less than 70 meters, i.e. 230 feet! Even
allowing very generously for exaggeration or poor estimation of
length, this creature was clearly of very considerable size.
Alternatively, part of what seemed to be the creature may in reality have been its wake, or perhaps the creature was actually a composite—several smaller snakes swimming in a line. When the
officer described their sighting to a monk, however, the latter told

him that it was indeed a *naga*.

Back in 1992, workmen attempted to demolish an old dilapidated abandoned temple in Phon Pisai, planning to replace it with a new one. Each time they approached this temple, however, a huge black thick-bodied snake partially emerged from it and reared up at them. Not only the workmen but also the village's Buddhist abbot saw it clearly. Only after an offering was left for it did the *naga* finally depart, enabling the workmen to demolish the old temple and replace it with the ornate modern version standing there today. Fable or fact?

Hope of examining tangible evidence for the *naga's* existence was raised when Richard learned of a purported *naga* bone, preserved inside a silver chalice as a holy relic and retained in Phon Pisai. Disappointingly, however, when he was granted permission to view the bone and the chalice was opened, all that it contained was a clearly identifiable elephant tooth!

Nevertheless, Richard believes that the *naga* may have some basis in reality. First evolving in the Cretaceous Period, the madtsoids were a group of primitive but huge snakes that were primarily aquatic, killed by constriction, and, in certain areas of the world, such as Australia, survived until only ten or so millennia ago. The most famous mythical entity of Australian aboriginal Dream-Time belief is the giant rainbow serpent. Although traditionally deemed to be fictitious, it may have been based upon a real creature after all.

In 2000, Queensland University zoologist Dr. Michael Lee, Dr. John Scanlon from New South Wales University, and colleagues proposed that the rainbow serpent might have been inspired by preserved memories of bygone encounters between early aboriginal settlers Down Under and a now-extinct genus of Aussie reptile comprising two species of extremely sizeable python-like madtsoid, the wonambi. Measuring 10 to 16 feet long, with a cross-section the size of a large dinner plate, it would have been big enough to swallow very large creatures, and thus could have readily given rise to legends of immense, all-consuming rainbow serpents. So perhaps the wonambi did survive long enough to have coexisted for a time with the first humans to reach Australia. And what of the Thai *naga*—could this be a living, modern-day Asian madtsoid?

Lastly, but no less intriguingly, Myanmar (formerly Burma) may have its very own version of the *naga*. Zoologist Dr. Alan Rabinowitz's recent book *Beyond the Last Village* (2001), featuring his explorations of this country, contains some tantalizing details of a hitherto-unpublicized mystery reptile:

> In Putao [a northern Burmese village], there were stories about a giant water snake, the bu-rin, 40 to 50 feet long, that attacked swimmers or even small boats. Sounding somewhat like a larger, aquatic version of the Myanmar python, this snake was considered incredibly hostile and dangerous. No one had firsthand knowledge of the creature, yet because of it, children were often discouraged from spending long periods of time in the water.

Is the *bu-rin* merely a fictitious, serpentine bugbear utilized by parents to frighten their children away from dangerous stretches of water, or could it simply be an extra-large, or exaggerated, normal python? Or is it something very different—an undisclosed living fossil? Perhaps yet another slithery surprise is lurking in a far-off region of the world, resolutely unknown to science but only too well known to the local people sharing its remote terrain.

CHAPTER 11

Exposing the
Chinese Ink Monkey

The galagos, or bush-babies, would materialize magically, like fairies, and sit among the branches, peering about them with their great saucer-shaped eyes, and their little incredibly human-looking hands held up in horror, like a flock of pixies who had just discovered that the world was a sinful place.

GERALD DURRELL—*ENCOUNTERS WITH ANIMALS*

THE VU QUANG OX *PSEUDORYX NGHETINHENSIS*, the giant muntjac *Megamuntiacus vuquangensis*, the Taingu civet *Viverra tainguensis*, and the Panay cloud rat *Crateromys heaneyi* are just a few of the remarkable number of major new mammals described from Asia during the 1990s. Yet none has a more controversial pedigree than the very mysterious mini-mammal supposedly rediscovered during the mid-1990s in southeastern China.

The tangled tale of the Chinese ink monkey (also called the pen monkey) hit the media headlines on April 22, 1996, when the *People's Daily* newspaper published an account based upon information supplied to it by the official New China News Agency (Xinhua). According to these sources, this amazing little animal was no larger than a mouse, weighed only seven ounces, and had recently been discovered alive in the Wuyi Mountains of Fujian Province after having been dismissed as extinct by Chinese scientists for several centuries. Strangely, however, no additional details concerning its resurrection—precisely when, by whom, the number of speci-mens recorded, whether any had been photographed or collected, etc.—were given. Nor was its zoological species identified, or any extensive morphological description supplied.

In contrast, details of its "pre-extinction" history were as pro-fuse as they were perplexing. Supposedly known in China since at least 2000 BC, the ink monkey was so named because it was the tra-ditional pet of scholars and scribes—and for good reason. Despite its tiny size, the ink monkey was very highly intelligent—so much so that a writer would frequently train one to prepare his ink for him, turn the pages of a manuscript that he was working on or read-ing, and pass brushes to him as required. Ever the practical, diligent pet, a writer's ink monkey would even sleep in his master's desk drawer or brush pot. Zhu Xi (AD 1130-1200), a famous Neo-Confucian philosopher, allegedly owned one of these obliging lit-tle helpers, yet sometime later the species seemed to have become extinct (an unexpected occurrence for so useful a beast, I would have thought) until its belated reappearance during the mid-1990s.

Those, then, were the "facts" of a very odd cryptozoological case—but they served only to enhance rather than to elucidate the mysteries encompassing this extraordinary animal.

First and foremost of these was its identity—what exactly *was*—or *is*—the ink monkey? The world's smallest known living pri-mate is itself a recent revival, the western rufous mouse lemur *Microcebus myoxinus* of Madagascar, measuring a mere 7.75 inches long, and weighing just over one ounce. This tiny species was first described by science back in 1852, but was subsequently believed to have died out until its rediscovery in 1993. Could it be that the ink monkey report was a confused retelling of this lemur's reap-pearance? In view of the historical background details presented above, this seemed highly unlikely.

Turning from lemurs to monkeys, however, shed little light on the problem either. The world's smallest formally recognized species of monkey is the pygmy marmoset *Cebuella pygmaea* of the

Upper Amazon Basin, with a total length of around 10 inches. Yet mainstream zoology knows of no species of Chinese monkey as small as a mouse, nor, as far as I am aware, are there any reports of such a beast in the cryptozoological literature either. Nevertheless, the ink monkey is not entirely unknown to Westerners.

I am greatly indebted to Steve Moore, a renowned specialist in Oriental Fortean-related literature, for bringing the following excerpt to my attention, which appeared in a book of Chinese lore by E.D. Edwards, *The Dragon Book* (1938), page 149:

> THE INK-MONKEY
> This creature is common in the northern regions
> and is about four or five inches long; it is
> endowed with an unusual instinct; its eyes are like
> cornelian stones, and its hair is jet black, sleek and
> flexible, as soft as a pillow. It is very fond of eating
> thick Chinese ink, and whenever people write, it
> sits with folded hands and crossed legs, waiting
> till the writing is finished, when it drinks up the
> remainder of the ink; which done, it squats down
> as before, and does not frisk about unnecessarily.

Edwards obtained these engrossing snippets from a Chinese author called Wang Ta-Hai, writing in 1791, whose work was translated into English as a two-volume tome titled *The Chinese Miscellany* (1845, 1849).

Whereas Edwards's description may initially conjure forth bizarre images of a red-eyed pixie with an insatiable ink lust, it also calls to mind certain real-life creatures, such as the bushbabies or galagos, and, especially, those real-life goblins of the golden eyes, the tarsiers. These tiny arboreal primates weigh only a few ounces and have brown or grey bodies measuring no more than six inches. Tarsiers are distantly related to bushbabies and lemurs, and are characterized by their large ears, enormous orb-like eyes, very long tarsal bones, flattened sucker-like discs at the tips of their fingers and toes, and a long thin tail.

TARSIERS

Traditionally comprising a trio of species indigenous to Borneo, Sumatra, the Philippines, and Sulawesi, in recent times at least three more species have been recognised from Sulawesi and its offshore islets—the pygmy tarsier *Tarsius pumilus* in 1987, Diana's tarsier *T. dianae* in 1988, and the Sangir tarsier *T. sangirensis* in January 1996. Could the ink monkey be yet another lately revealed tarsier? During a radio interview broadcast on April 24, 1996, concerning the ink monkey in which he expressed the view that it may possibly be a tarsier, veteran British wildlife broadcaster Sir David Attenborough noted that tarsiers did exist in southern China a long time ago, and that such secretive, nocturnal creatures as these might conceivably have succeeded in surviving in small numbers amid this region's thick forests right up to the present day while remaining undetected by science.

Conversely, during the same interview, Cyril Rosen of the International Primate Protection League favored the slow loris *Nycticebus coucang* as a plausible contender, whose southeast Asian distribution range may indeed extend into southeastern China.

Yet even if one or other of these identities were correct, how could such an animal be trained to prepare ink? Back in the centuries when the ink monkey allegedly assisted writers in this manner, ink was normally compounded from a precise array of precious materials, such as sandalwood, musk, gold, pearls, and rare herbs or bark, yielding a stick of ink. Consequently, it was suggested in Western media accounts that perhaps the ink monkey was trained to grind the stick in an inkstone with pure water until the correct shade was obtained. Nevertheless, such valuable constituents as those listed above are hardly the types of material that most people would entrust to a monkey (or tarsier) for handling.

Moreover, aside from Edwards's curious contribution that the ink monkey not only prepares but also consumes ink, there were no details as to its dietary preferences. Tarsiers, unlike most other primates, are entirely carnivorous. So if the ink monkey is indeed a tarsier, what might we expect its favorite food to be? When a colleague's enquiries at the news agency in Xinhua that released the original press information concerning the ink monkey revealed that the agency had no record of where they had actually received their own details(!), I began to suspect that fillet of red herring—or perhaps even a tasty canard?—may well prove to be the answer.

In other words, it seemed likely that the whole episode of the Chinese ink monkey was either a bizarre hoax from start to finish, or, more probably, simply the benighted product of confused reporting. Quite possibly, for instance, a Chinese media report had somehow been erroneously translated by Western journalists so that its true subject, one of China's lesser-known non-existent beasts of fable and folklore, had falsely "become" a real-life animal that had been rediscovered after several centuries of supposed extinction.

And in case this hypothesis seems too implausible, I recommend readers to peruse an entire chapter devoted to several comparable cases of monstrous misidentification exposed in all their glorious folly, within one of my earlier books, *From Flying Toads to Snakes With Wings* (1997).

Nevertheless, it would be reassuring if the original source of such confusion could be traced. For a time, I was unable to do this, but eventually, when resurveying zoological reports from April 1996, I came upon a short piece that had not attracted anywhere near as much attention as the Chinese ink monkey story, and which I had not seen before, yet which assuredly presented me with the missing piece of this extraordinary cryptozoological jigsaw.

The report was a short but succinct item published by London's *Times* newspaper on April 5, 1996 (three weeks before the ink monkey reports). It recorded the discovery in China's Yellow River basin of fossil jaw remains from a mouse-sized primate dubbed *Eosimias centennicus*, which had lived 40 to 43 million years ago, and weighed a mere three to four ounces. Needless to say, this description tantalizingly echoes the sparse details available in the subsequent reports dealing with the Chinese ink monkey.

Consequently, the most reasonable explanation for this entire episode is as follows. Somewhere in the early portion of the chain of journalistic communication ultimately giving rise to the ink monkey media stories in the West, details of the discovery of the *Eosimias* fossils, documented correctly in *The Times*, became distorted elsewhere, and were mistakenly combined with the traditional Chinese ink monkey folklore—until the tragi-comical result was the rediscovery of a creature that had never existed.

In other words, a scenario featuring, perhaps fittingly, a classic case of Chinese whispers!

CHAPTER 12

Living Unicorns and Latter-Day Dragons—The Physical Reality of Fabulous Beasts

The unicorn was white, with hoofs of silver and graceful horn of pearl. He stepped daintily over the heather, scarcely seeming to press it with his airy trot, and the wind made waves in his long mane which had been freshly combed. The glorious thing about him was his eye. There was a faint bluish furrow down each side of his nose, and this led up to the eye-sockets, and surrounded them in a pensive shade. The eyes, circled by this sad and beautiful darkness, were so sorrowful, lonely, gentle and nobly tragic, that they killed all other emotion except love.

T.H. WHITE—*THE ONCE AND FUTURE KING*

Be assured the Dragon is not dead
But once more from the pools of peace
Shall rear his fabulous green head.

ROBERT GRAVES

THE POSSIBILITY THAT IMAGINARY, FABULOUS ANIMALS of traditional myths and legends have a firm foundation in fact may well seem like a contradiction in terms. As will be revealed here, however, there is all manner of tantalizing evidence and clues on file pointing to the erstwhile physical reality of this elusive, and illusive, fauna of reverie and folklore.

THE UNICORN—DISPERSING THE MAGIC OF HORN AND HOOF

A fearless yet fundamentally tranquil beast, famously reclusive but readily enticed and tamed by the presence of a virginal maiden, and equipped with a spiraled horn capable of exposing and nullifying the most toxic poisons, the unicorn is unquestionably the most celebrated and ethereal of the world's panoply of fabulous animals. It is also one of the most controversial—due to the surprising diversity of animals allegedly contributing to its origin.

In the fourth century BC, the Greek physician Ctesias recorded the supposed existence in India of a wonderful creature resembling a white ass with blue eyes, and a dark red head whose brow bore an 18-inch-long horn with a pure-white base, jet-black stem, and crimson tip. This variety of unicorn was probably derived from reports of the onager or Asian wild ass *Equus hemionus onager,* combined with distorted accounts of the single-horned great Indian rhinoceros *Rhinoceros unicornis.*

Very different was the white, antelope-like version with cloven hooves supposedly spied at Mecca in 1503 by traveler Vertomannus; this was certainly the Arabian oryx *Oryx leucoryx,* which possesses a pair of long straight horns, and closely resembles a unicorn when observed in profile. Reports of Tibetan unicorns were also based upon an antelope, the chiru *Pantholops hodgsoni;* and Nepal's equivalents were domestic rams whose horns had been artificially fused into a single, centrally sited horn.

The biblical unicorn or *re'em* was actually the European wild ox (aurochs) *Bos primigenius,* a two-horned species that became extinct in 1627. However, in 1934 Maine University biologist Dr. W. Franklin Dove succeeded in creating a "genuine" bovine unicorn—by removing the embryonic horn buds from a day-old Ayrshire bull calf, trimming their edges flat, and then transplanting them side by side onto the center of the calf's brow. Growing in close contact with one another, they yielded a massive single horn, which proved so successful a weapon that its owner soon became the undisputed leader of an entire herd of cattle. Yet despite his dominance, this unicorn bull was a very placid beast, thus resembling the legendary unicorn not only morphologically but also behaviorally. Coincidence? Perhaps ancient people knew of this simple technique and had created single-horned herd leaders, whose imposing appearance and noble temperament thereafter became incorporated into the evolving unicorn legend.

As for the many unicorn horns (termed alicorns) prized over the years, these have an even more unexpected identity—the ornate, spiralled tusk of the male narwhal *Monodon monoceros*, an unusual species of whale, whose highly prized tusks were obtained when narwhals were harpooned by whalers or when dead narwhals were beached. Washed ashore on Frobisher's Bay, Canada, during the 16th century, and presented to Queen Elizabeth I, one of the most famous alicorns is the seven-foot-long specimen, ornately tipped with silver that forms part of the English Royal Family's crown jewels.

In April 1700, numerous supposed "unicorn bones" were disinterred from a clay pit on the River Neckar's bank, near Cannstatt, Switzerland, and were taken to Duke Eberhard Ludwig, who donated some to the City of Zurich. Their long-deceased owner became known as the Schwabian unicorn, but paleontologists eventually revealed that its bones were actually those of a heterogeneous assortment of fossil species (including lions), and its horn was a mammoth's tusk.

In November 2002, Auckland Harbour customs officers confiscated from an Auckland furniture importer an undeclared crate of ostensibly remarkable items. For the crate, shipped in by him from Indonesia, was found to contain bones that the furniture importer genuinely believed to be from a unicorn! He had brought them, apparently in good faith regarding their identity, as a special Christmas present. When the bones were examined at Auckland Museum, however, they were identified as originating from either a cow or a water buffalo, and had been coated with a thin layer of cement to make them appear fossilized.

Speaking of fossils: it is possible that there was once a bona fide species of fossil unicorn. Approximately one million years ago, a remarkable antelope lived in Europe, known as *Procamptoceras brivatense*. Now extinct, it was notable for the extreme nearness to one another of its two upward-pointing, in-line horns. Indeed, as pointed out by renowned paleontologist Prof. Bjorn Kurten in his book *The Ice Age* (1972), in life this species' horn-cores were covered with horny sheaths "that must have been so closely appressed that the animal looked like a unicorn." Although current fossil evidence suggests that *Procamptoceras* died out before humans reached Europe, if this species did linger here into more recent times and became known to humans, who could say with certainty that it did not influence our ancestors' development of the unicorn myth?

Perhaps the most fascinating unicorn link of all, however, follows in the wake of an extraordinary recent discovery. In 1992, a radically new species of animal, hitherto wholly unknown to science, was revealed in Vietnam. Now known as the Vu Quang ox *Pseudoryx nghetinhensis*, it is a spindly legged relative of the Asian wild oxen, but with long oryx-like horns that blend into one when the creature is viewed side-on. Probably the earliest form of unicorn is the Orient's *ki-lin*, reports of which date back as far as 3000 BC. Yet the *ki-lin's* zoological inspiration has never been satisfactorily explained, because Asia did not appear to contain any living species remotely resembling it. As suggested in October 1993 by P.J. Sharp from Edinburgh, however, perhaps it was based upon the Vu Quang ox. How ironic it would be if a "new" animal proved to be the identity of one of the most ancient animals.

GRIFFIN, OR GRIFFINOSAURUS?

Another ancient beast of myth and legend is the griffin—a fabulous monster with the body of a lion but the wings and head of an eagle. It apparently originated in the Levant region, with accounts of griffins from this area dating back more than 4000 years, but later spreading west into Europe. They were well known for their fondness for gold—Greek historian Herodotus, writing in the fifth century BC, stated that griffins inhabited the lofty mountain peaks of India, where they dug up gold with which to construct their mighty nests, which in turn were greatly coveted by opportunist gold-hunters. Also highly sought after were their long curved talons, allegedly capable of detecting poison, and many were brought back to Europe during the Middle Ages by crusaders. Sadly, however, they invariably proved to be antelope horns, sold to the crusaders by African entrepreneurs!

GRIFFIN

Traditionally, zoologists have sought to identify the griffin with a powerful species of vulturine bird of prey called the lammergeier *Gypaetus barbatus;* but in recent times, as with the unicorn, a very novel alternative candidate has been proposed. In 1991, Dr. Adrienne Mayor, a classical scholar from Princeton, New Jersey, noted that the Altai Mountains of central Asia, famous for their rich gold deposits and the locality of many ancient griffin legends, also contain the fossilized skeletons and eggs of a lion-sized quadrupedal dinosaur called *Protoceratops,* whose most distinctive feature is its strikingly eagle-like beaked head. Dating from the Cretaceous Period (136 million to 64 million years ago), if its remains had been encountered by early gold prospectors many centuries before the correct identity of dinosaurs was recognized, how would they have been explained by such people? Quite possibly as the skeletons of four-footed eagle-headed monsters—or, as we call them today, griffins.

SENMURV, SIMURGH, AND A ONCE-LIVING ROC LEGEND

Closely allied (if not directly ancestral) to the griffin, and sometimes termed the cynogriffin, is the senmurv ("dog-bird") of early Persian mythology, which combines the body of a lion, and the wings and talons of an eagle, with the head of a dog. Its home was a magical tree, whose seeds, scattered throughout the world whenever the senmurv alighted in its mighty branches, cured diseases and evils of every kind. Later Persian legends transformed the senmurv into a huge, resplendent bird known as the simurgh, which nested on the highest peak of northern Persia's Alburz Mountains, and was permanently hidden from mortal view behind a shimmering veil of darkness and light. Nevertheless, this supposedly unseen bird was frequently, and lavishly, portrayed in ancient bestiaries and manuscripts—a paradoxical situation that incited one short-tempered scholar long ago to pen the following terse but apt comment in the margin of an exquisite plate depicting the simurgh in a 13th century bestiary: "Thou fool, if nobody has seen the simurgh, then how dost thou portray it?"

Folklorists consider it most likely that the simurgh was based upon one or more of the large vultures and eagles native to Iran and elsewhere in the Middle East, and its "evolution" was probably also influenced by Arabian myths of the enormous roc or rukh, a gargantuan bird of prey that could carry elephants aloft in its huge talons and inhabited an unspecified island in the Indian Ocean. Fronds from the exceptionally large palm tree *Sagus ruffia* were once believed to be the feathers of the roc. Moreover, giant subfossil eggs of the now-extinct flightless elephant bird *Aepyornis maximus*, shown to European sailors when they first began to visit the island of Madagascar a few centuries ago and able to hold two gallons of fluid, were originally assumed to be roc eggs. And indeed, it is certainly possible, notwithstanding its flightlessness, that the elephant bird, which stood over nine feet tall and may have survived until the late 1600s, inspired early tales of the imaginary roc.

HERE BE DRAGONS—FROM STUFFED LINDORMS TO FOSSILIZED DRAGONETS

No mythical animal is more famous (infamous?) or more widespread among the world's legends and folklore than the ferocious dragon, in all its many diverse forms and guises. However, humanity's erstwhile belief in such monsters was probably inspired at least in part by encounters with real animals, such as giant monitor lizards like Salvadori's monitor *Varanus salvadorii* from New Guinea, which can attain lengths of up to 15 feet, or the slightly shorter but much bulkier Komodo dragon *V. komodoensis*, whose ever-flicking vivid red tongue could have been readily transformed by the boundless talents of human imagination into flickering flames of fire. Some large lizards, moreover, are semi-aquatic, and these, together with crocodiles, might well have inspired fables of amphibious water dragons.

Also, in bygone, pre-paleontological days, uneducated people encountering the fossilized remains of dinosaurs and other pre-historic animals erroneously assumed that they were the mortal remains of dragons. There is little doubt, for example, that medieval legends of a long-necked, long-winged dragonet inhabiting Mount Pilatus in Switzerland drew their inspiration from the numerous fossilized pterodactyls discovered in this mountain's rocks.

Fossilized dinosaur remains in Asia have also been proposed as the origin of the ancient Eastern belief in dragons. Back in 1916, J. O'Malley Irwin suggested that the spectacular remains of long-necked sauropod dinosaurs such as *Camarasaurus (=Morosaurus)* discovered by him and others in China may well have inspired this land's traditional dragon legends. And the remains of fossil human teeth here were commonly thought to be dragon teeth.

In his book *The Dragons of Eden* (1977), eminent Cornell University scientist Prof. Carl Sagan contemplated a still closer link between fictional dragons and factual giant reptiles, speculating that perhaps some prehistoric reptiles survived into more recent times than currently accepted by science, and that mankind's myths and legends of dragons may therefore stem from ancient, inherited memories of our long-distant ancestors' encounters with *living dinosaurs*.

Whereas many continental European dragons were of the classical winged, quadrupedal, fire-disgorging type, most British dragons are of the worm variety, with lengthy, elongate bodies, but limbless, wingless, and emitting poisonous vapors rather than fire. One of the most famous of these vermiform horrors was described as follows:

> Ane hydeous monster in the forme of a
> Worme...in length three Scots yards and some-
> what bigger than an ordinary man's leg, with a
> head more proportionable to its length than to its
> greatness, in forme and colour like to our com-
> mon muir adders.

This is a medieval description of the Linton worm, a monstrous creature that issued forth from a tributary of Scotland's River Tweed in the 12th century, to begin a reign of terror within the surrounding countryside. Making a den for itself on Linton Hill, in Roxburghshire, it preyed upon the farmers' livestock, asphyxiating them with its poisonous breath and sunbathing upon the hill after a heavy meal. In desperation, the local populace offered a huge reward to anyone who would free them from this loathsome beast, and in 1174 their plea was answered by a valiant nobleman, the Laird of Lariston, believed to be a member of the Somerville family.

In the Apocrypha, there is a story of how Daniel killed a dragon by forcing lumps of pitch, fat, and hair into its mouth. The Laird of Lariston adopted a similar means of dispatching the Linton worm, by thrusting down the creature's throat a lump of burning peat dipped in red-hot sulphur and pitch, attached to the tip of his lance. This dynamic scene is portrayed in a handsomely sculptured tympanum in Linton's parish church.

Many of Britain's dragon tales are believed to be wholly fictitious, but there may be at least a core of fact at the base of the Linton worm's legend. The description of its appearance recalls a very large snake, as does its sunbathing behavior following a meal, and its poisonous breath (like other carnivorous animals, large snakes have foul-smelling breath). It could well have been a python or some other exotic big snake that had escaped from one of the many menageries popular in medieval times. Such cases are far from unknown.

The Wormingford "corkindrill" killed in 1405 on the border of Essex and Suffolk, for instance, was certainly an escapee crocodile from the Tower of London, where it had been maintained since its arrival in England from the Crusades.

In Europe, limbed but wingless dragons were known as lindorms, and several preserved or stuffed specimens of supposed lindorms have been documented over the years. The journal *Notes & Queries* (December 28, 1850, and January 18, 1851), for instance, mentions a Moravian example suspended in an arched passage leading to Brünn's town hall; another one hanging from the roof of Abbeville Cathedral, Picardy; and also an example in the church of St. Maria delle Grazie, near Mantua, northern Italy. Such exhibits are generally large lizards or—in the case of the Moravian specimen—crocodiles. However, the "lindorm skull" discovered at Klagenfurt, southern Austria, after being found here in 1335, and which was the inspiration for this town's impressive sculptured fountain of a lindorm, completed in the late 16th century, was unmasked by zoologist R. Pusching in 1935 as that of a woolly rhinoceros!

GESNER'S SEA-MONK (LEFT) AND SEA-BISHOP (RIGHT)

THE BEASTS THAT HIDE FROM MAN

PSEUDO-SEA MONSTERS AND THE
HYDRA OF HERACLES

It is well known that the many specimens of hideously ugly "taxi-derm mermaids" in existence have been skillfully manufactured by sailors and others seeking to part gullible travelers from their money by sewing together the upper parts of monkeys and the lower parts of fish.

Even more plentiful, however, and equally bizarre in appearance is a multifarious assemblage of preserved marine monstrosities variously dubbed sea-monks, devil-fishes, and Jenny Hanivers. In medieval times, their reality was widely accepted, and hence they were frequently portrayed in bestiaries, alongside sea life of a much more reputable nature.

Thus, on page 174 of Conrad Gesner's *Nomenclator Aquatilium Animantium* (1560), we find two extraordinary engravings. One is of a sea-monk, supposedly caught off the coast of Norway during Gesner's own lifetime, but which bears a recognizable if somewhat distorted likeness to a large squid (giant squids have often been washed ashore around Scandinavia).

The other is of a so-called sea-bishop, which was allegedly spied in 1531 off the coast of Poland. This too could conceivably be a rather imaginative illustration of a squid, and was possibly the engaging if erroneous outcome of an artist attempting to illustrate a creature known to him only from second-hand descriptions.

Moreover, judging from various suspiciously octopus-like renditions on ancient Greek vases and other depictions from that age and locality, one of Heracles's most formidable opponents, the multi-headed hydra, was either a monstrous octopus or a colossal squid.

And in a fascinating article published by the American magazine *Natural History* in June 1941, cryptozoological writer Willy Ley put forward an interesting argument in favor of accepting Homer's description of another, equally terrifying monster of Greek mythology—Scylla, a hideous sea-dwelling entity with six heads on six long necks, and twelve misshapen feet—as the earliest documentation of a giant squid.

Even more grotesque than Scylla, however, is an engraving on page 166 of Gesner's aforementioned bestiary, portraying an Indian sea monster that was somehow found in Milan. And on page 33 of the supplement (named *Paralipomena*) of Gesner's *Historia Animalium Liber IV. Qui est de Piscium et Aquatilium Animantium Natura*, 2nd edition (1604), there can be found an engraving of a truly hideous monstrosity perversely labeled a sea-dove. Looking at both of these freakish fish, however, it is not difficult to discern that they are nothing more than artfully modified, dried specimens of rays or skates. So too are devil-fishes, which can still be bought in curio shops in many a seaside resort in Europe, and also in America, where, keeping up with the times, they are often sold to unsuspecting tourists as the earthly remains of dead extraterrestrials. But that, as they say, is another story!

Most elaborate of all are Jenny Hanivers. Works of art in their own right, these are rarely seen nowadays, but once again originated as dead rays or skates, whose broad lateral fins had been carefully cut, molded, and dried into the shape of curving limbs, thus creating all manner of realistic if decidedly fishy (in every sense!) forms of water dragon.

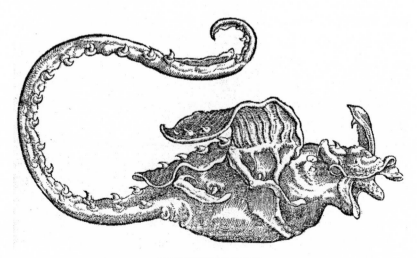

GESNER'S SEA-DOVE

THE BEASTS THAT HIDE FROM MAN

THE BONES OF THE THUNDER HORSE

Perhaps the most surprising relics of a mythical beast, however, are those of the thunder horse. According to Oglala Sioux tradition in Nebraska, Wyoming, and South Dakota, this is a huge, terrifying creature that leaps down from the skies to the earth during storms, and whose sonorous hoofbeats are the origin of thunder. Predictably, scientists initially discounted this as fantasy, until 1875, when the Sioux showed American paleontologist Prof. Othniel C. Marsh some enormous bones that they claimed were from a thunder horse, and which were unlike those of any animal species known at that time! Greatly intrigued, Marsh studied them, and revealed that they were the fossilized remains of a hitherto unrecorded rhino-like relative of horses that had lived about 35 million years ago. Commemorating its links with legend, Marsh dubbed it *Brontotherium*—"thunder beast"—the most famous member today of a totally extinct family of ungulates known as titanotheres.

BIRDS OF PARADISE AND THE PHOENIX LEGEND

Everyone is familiar with the age-old legend of the phoenix, in which this exquisitely plumaged Middle Eastern bird, the only one of its kind, dances in the midst of its burning nest until it is entirely consumed by the flames, yet is afterwards miraculously resurrected. But did the phoenix really exist, and how did its extraordinary legend originate?

For centuries, scientists have sought to identify the phoenix with a real bird, such as the peacock, golden pheasant, golden eagle, purple heron, various parrots, and even one or more of New Guinea's crow-related birds of paradise. This last-mentioned possibility has attracted especial interest, as there are several different birds of paradise that correspond very closely to ancient descriptions of the phoenix's reputed appearance.

Morphological similarity, however, is not sufficient in itself. After all, how could a bird of Middle East mythology have been based upon any species native to a land as far away from Asia

Minor as New Guinea? This remained a seemingly insoluble problem, until 1957, when a team of Australian zoologists, investigating the bird of paradise skin trade in New Guinea, made some extraordinary discoveries.

They learned that although Western science only became aware of these spectacularly beautiful birds' existence in the 16th century AD, some New Guinea tribes had been exporting their exquisitely plumed skins for untold centuries before this—as far back, in fact, as 1000 BC, when this island was visited by Middle Eastern seafarers from Phoenicia, home of the phoenix legend! The team also learned that to ensure that the fragile skins remained intact during the long, arduous sea voyage back home from New Guinea to Phoenicia, the tribespeople used to wrap them in myrrh, molded into an egg-shaped parcel, and then sealed this inside a second parcel composed of burned banana leaves.

This vividly recalls the phoenix myth, in which the reborn phoenix preserves its nest's ashes in a package of myrrh, wrapped in turn within aromatic leaves fashioned into the shape of an egg, which it then leaves as a tribute on the altar of the sun god's temple at Heliopolis. The most startling correspondence between myth and reality, however, is still to come.

To obtain the highest prices for these skins, the tribespeople would naturally have ensured that they were from the most sumptuously plumaged species, and it just so happens that one of these is also one of the most abundant—thereby guaranteeing its plentiful representation in any consignment of skins for export. This species is Count Raggi's bird of paradise *Paradisea raggiana*, the male of which is adorned with two lavishly profuse sprays of bright scarlet feathers cascading flamboyantly from beneath its wings. During its premating display, it performs a sensational courtship dance, quivering animatedly and repeatedly elevating its fiery plumes; to a human observer, the visual effect is uncannily like a bird dancing amid a blazing inferno of flame! Descriptions of this extraordinary spectacle retold back home in Phoenicia by maritime voyagers returning from New Guinea, and substantiated by the egg-molded parcels of skins from the exotic species performing this dramatic ritual, would have been more than adequate to inspire the phoenix legend.

Moreover, the celebrated Russian firebird may also have been based upon birds of paradise. In 1961, Soviet scientist V. Kiparsky discussed this thought-provoking concept when contemplating the possibility that their skins had first reached Eastern Europe far earlier in history than previously assumed.

THE CELESTIAL TIGERS OF ANNAM

According to the traditional mythology of the Annamite people in central Vietnam, the four corners of Space are ruled by four celestial tigers. The Black Tiger is the ruler of the North, and his empire incorporates winter and water. The Red Tiger rules the South, and governs summer and fire. The White Tiger, ruler of the West, controls autumn and metals. And the Blue Tiger, reigning in the East, is the monarch of spring and plants.

Remarkably, all four of these highly unusual tigers may well have a basis in reality. It is well known that several species of wild cat sometimes produce freak, all-black (melanistic) specimens, due to various mutant genes. The black panther, for instance, is merely a melanistic mutant of the leopard, and melanistic jaguars are also very common. Moreover, although no specimen has so far been made available for formal scientific examination, there are many reports by reliable, experienced eyewitnesses on file of all-black tigers, and in the 1970s, a partially melanistic specimen was born at Oklahoma City Zoo. Genetically, there is no good reason why melanistic tigers should not arise occasionally, and it is quite conceivable that such distinctive specimens inspired the Annamese belief in the celestial Black Tiger.

Similarly, a number of naturalists and other skilled eyewitnesses have reported seeing in the Sunderbans area of eastern India a strain of tiger whose stripes are barely visible against their brownish-red coats, so that they appear stripeless. Once again, such eyecatching specimens as these might well be sufficient to engender the legendary celestial Red Tiger.

As for white tigers: the modern-day strain commonly seen in zoos and wildlife parks is largely if not entirely descended from a magnificent male white tiger called Mohan, captured alive in Rewa on May 27, 1951, and later housed within the summer palace of the Maharajah. However, white tigers have been reported elsewhere in India too, as well as in China, Korea, and Nepal, and the earliest known record of white tigers dates back to 1561, when two of the cubs of a tigress killed by the Mogul emperor Akbar were reported to be white. Consequently, the occurrence of white tigers in Eastern mythology is hardly unexpected.

What may well be unexpected, conversely, is the prospect that even the most dreamlike of the four celestial tigers, the Blue Tiger, could well have a living counterpart. For countless years, there have been reports of blue tigers inhabiting Korea, and particularly Fujian Province in southeastern China, occasionally spied by local people but never captured or shot. In September 1910, however, their reputation for elusiveness suffered a serious blow, when one of these ethereal beasts was almost bagged by a keen American game hunter.

Harry R. Caldwell, who was also a Methodist missionary, was hunting in Fujian when one of his native helpers suddenly directed his attention toward something moving in the undergrowth nearby. Caldwell thought at first that it was another native, dressed in the familiar blue garment worn by many in that region, but when he peered more closely he realized to his great surprise that he was actually looking at the chest and belly of a tiger—a very large tiger whose black-striped fur was not orange-brown, as in normal specimens, but was instead a very distinctive shade of blue.

Caldwell decided to shoot this extraordinary creature, to prove once and for all that such cats really did exist, but the tiger was watching two children gathering vegetation in a nearby ravine, and he knew that if he tried to shoot it from his present position he might injure or otherwise endanger the children too. Consequently, he moved a short distance away, in order to alter the direction of his planned shot, but while he was doing this, the tiger disappeared into the forest, and was not seen by him again.

Although a blue tiger seems impossible, in fact it is quite easily explained. Such tigers almost certainly possess the genes responsible for the smoky blue-mauve coloration characterizing the Maltese breed of domestic cat; in America, fur traders have occasionally obtained comparably hued specimens of the lynx and the bobcat.

FATHER HEN AND THE COCKATRICE

According to traditional European mythology, if a cockerel lays a leathery shell-less egg that is hatched in a dung-heap at midnight by a toad, the resulting offspring will be a cockatrice—a monstrous crowing beast combining a scaly reptilian body with the feathered neck, legs, sharp beak, and coxcombed head of a cockerel.

Although it may all seem highly unlikely, the cockatrice's history does have a basis in fact, due to a bizarre fluke of nature whereby a hen will sometimes undergo a sex change. Through the ages, many such cases have been recorded (more than is widely realized, in fact), in which a seemingly normal hen abruptly begins to transform partially into a cockerel, growing male plumage and a coxcomb, and crowing each day, but often continuing to lay eggs.

In bygone, superstitious days, such a bizarre occurrence would have been seen as an omen of impending doom, leading in turn to the development of the cockatrice legend. In reality, however, the occurrence of one of these "father hens" can be sometimes due to an internal tumor that stimulates the hen to develop male hormones and thence acquire secondary sexual characteristics associated with cockerels. Alternatively, the hen may actually be a hermaphrodite—a freak specimen born with both female and male organs, in which the former dominate for the earlier part of its life but are later subordinated by the latter.

One well-publicized recent "father hen" is Henrietta, featured in a *Daily Mail* report on March 4, 1998, who is (or was) one of several hens on Violet Smith's farm at Aylmerton, near Cromer, Norfolk. For ten years, Henrietta was a totally normal hen—until 1997, that is, when she began growing the green downward-pointing tail feathers, the shiny russet neck plumage, and the large red coxcomb of a typical cockerel. By Christmas 1997, much to the consternation of the farmyard's real cockerel, she had taken charge of the other, normal hens, and during the last week of February 1998 she gave voice to her first cock-crow. Not surprisingly, she was duly renamed Henry, and has since stopped laying eggs, but experts believe that she may well begin again, just like other father hens often do—but if so, it may be advisable to keep these eggs away from toads and dung-heaps, just in case!

The history of fabulous animals is as complex as many of the animals themselves—deftly but intricately interweaving the morphology and behavior of real creatures with the ever-inquisitive observations of these beasts by our far-distant ancestors, and their attempts to make sense of all that they encountered in the natural world. Yet although fuelled by the limitless powers and talent of the human imagination, in those bygone ages their attempts lacked the moderating control of scientific logic. Consequently, it should really come as no surprise to find that the outcome was a veritable menagerie of monsters—which were too fantastic ever to have lived, but which have also proved too amazing simply to be forgotten. And so these fabulous beasts have survived, animated not by nature but by the human mind, until their true origins have become obscured to all but their most tenacious researchers.

CHAPTER 13

Shamanus, Sun Dogs, and Other Canine Enigmas

There is a wolf in me...fangs pointed for tearing gashes...a red tongue for raw meat... and the hot lapping of blood — I keep this wolf because the wilderness gave it to me and the wilderness will not let it go.

CARL SANDBURG—*WILDERNESS*

ALL TOO OFTEN OVERSHADOWED BY THEIR MORE famous feline counterparts, canine mystery beasts are no less intriguing, or diverse, as the following selection of cases readily reveals. Whether they all involve genuine canids, conversely, is another matter entirely!

DOES JAPAN'S BONSAI WOLF STILL SURVIVE?

It is, perhaps, only fitting that Japan, famous for its miniature bonsai trees, should also lay claim to the world's smallest subspecies of wolf. Known as the shamanu or Japanese dwarf wolf *Canis lupus hodophilax*, and sporting a short, dense coat of predominantly ash-grey fur, it measured no more than 45 inches long, including its one-foot tail, and stood a diminutive 14 inches at the shoulder. This was due in particular to its disproportionately short legs, which morphologically alienated the shamanu so dramatically from all other wolves that some zoologists classified it as a wholly distinct species of canid in its own right.

In addition to the shamanu, Japan was once home to a "nor-

SHAMANU

mal" type of wolf too, known colloquially as the *yezo-okhami* and scientifically as *Canis lupus hattai*. However, this much larger subspecies was extinct by 1887, and had been restricted to Hokkaido.
In contrast, the vertically challenged shamanu, colloquially termed
the *nihon-okhami*, was not only present on Hokkaido, but also
thrived on Honshu, as well as the Kurile Islands.

Japan's aboriginal people, the Ainu, had referred to the
shamanu as the howling god, but its deified status was not enough
to save it from the bane of human persecution, which has been
inflicted upon wolves of all shapes and sizes throughout their geographical range. For decades, shamanus were feared, in spite of their
small size, and hunted accordingly. Further depletion resulted from
rabies, which first reached Japan from overseas in 1732, plus the
destruction of the shamanu's forest home for cedar replanting. And
so it was that whereas its bonsai trees are today flourishing not only
in their native land but also throughout the world, Japan's dwarf
wolves had been harried into extinction by 1905.

On January 23, 1905, Stanford University zoologist Malcolm Anderson, visiting Japan as part of a research team sent by London's Zoological Society and the British Museum, was offered for sale the recently killed corpse of a shamanu by three hunters at Washikaguchi, Nara Prefecture. For the lowly sum of eight yens 50 sens, he purchased it, and sent its skin to the British Museum, where it has remained ever since. Today, the Washikaguchi shamanu's value is incalculable, because it constitutes the last confirmed specimen on record.

But is the shamanu really extinct? Although it rarely rates a mention nowadays in Western zoological literature, there is a sizeable corpus of material in Japanese publications suggesting that this curious little canid may well have lingered on into the present day.

Probably the most extensive single source of such material is Yanai Kenji's book, *Visionary Japanese Wolves* (1993). Kenji was inspired to prepare the book by the previous lack of detailed, readily accessible information dealing with the shamanu, and, especially, by his own sighting, at the age of 48, of what he believed to be a surviving shamanu.

It was early morning on the first Sunday of June 1964 when, as an enthusiastic amateur mountaineer, Kenji was journeying with his son and a colleague to the Ryogami Mountain. After hearing a series of spine-chilling howls, the trio suddenly spied a mysterious canine creature that was as big as an alsatian. They were able to observe the animal closely for a few minutes before it eventually disappeared, and afterwards they discovered a half-eaten hare nearby that it had probably killed just before they had arrived on the scene.

Professor Fujiwara Shizuo, from Tokyo University's Department of Science, was convinced that Kenji had indeed seen a shamanu. Others, conversely, were more skeptical. Certainly, if the creature was the size of an alsatian, I find it difficult to believe that this could have been a shamanu. However, as documented in Kenji's book, he is not the only investigator to support the possibility of this distinctive canid's continuing survival.

Folklorist Yanagida Kunio, for instance, has written that the shamanu still existed as late as autumn 1934 in Oku-Yoshino, Nara Prefecture, a view also subscribed to by his assistant, Kishida Hideo. Moreover, Gokijo Yoshinori, who lived in the village of Shimo-Kitayama, Nara Prefecture, claimed that shamanus were still dwelling after World War II in the mountains stretching from the Odaigahara Heights to the Omine Heights in the Kii Peninsula. In 1970, journalist Hida Inosuke actually claimed to have photographed one of these mini-wolves at night in the Omine Heights—which, together with the Odaigahara Heights, is the locality offering the greatest probability of shamanu survival, in Kenji's opinion.

According to another shamanu investigator, Seko Tsutomu, an animal greatly resembling one of Japan's supposedly long-lost dwarf wolves was sighted by Saratani Takehiro on a cliff on the side of a dam in Nara Prefecture even more recently—in 1985.

Elsewhere in Japan, veterinary surgeon Oura Yutaka averred that in 1934, the shamanu still existed in the mountains bordering Fukushima, Yamagata, and Niigata Prefecture. And Kuno Masao drew a sketch of the shamanu that he allegedly spied in the Kumotori-Yama Mountain, situated on the border of Tokyo, Saitama, and Yamanashi Prefecture.

Not everyone, however, is convinced by this anecdotal evidence. Some Japanese zoologists believe that reports of shamanus merely derive from misidentifications of feral dogs. Certain others have wondered whether the creatures being seen might actually be hybrids, descended from original matings between pre-1905 shamanus and domestic dogs, but this proposed identity was ruled out by Dr. Hiraiwa Yonekichi, an eminent authority on the wolf.

Rather more tangible, yet no less tantalizing, is the preserved specimen of a shamanu that was discovered in January 1994 at a shrine in Kokufu-Machi, Tottori Prefecture—for it has been claimed, but not proven, that this specimen was presented to the shrine in 1950.

Nevertheless, its discovery revitalized public interest in the shamanu, which in turn led to the staging on March 19 and 20, 1994, of the 1st Japanese Wolf Forum, a two-day symposium held at Higashi-Yoshino's town hall in Nara Prefecture, and sponsored by the Nara Committee of Wild Life Conservation. As reported in a *Nihon Keizai Shimbun* newspaper report from March 20, 1994, about 80 shamanu researchers and other interested individuals from all over Japan participated in this event, which featured the testimony and recollections of hunters, mountaineers, and others with reports attesting to the shamanu's post-1905 persistence. Even a sample of alleged shamanu feces was presented by one investigator to the Committee, and following the extent of interest generated by the symposium, plans were drawn to entice shamanus into view by using imported Canadian wolves as lures.

In January 1997, a significant new piece of evidence for the shamanu's putative survival was publicized, a series of recent photographs depicting a dog-like beast that bears a startling similarity to a living shamanu. The photos were snapped on October 14, 1996, by 47-year-old Hiroshi Yagi in the mountain district of Chichibu. He had been driving a van along a forestry road that evening when a short-legged dog-like beast with a black-tipped tail suddenly appeared further down the road. A longstanding shamanu seeker, Yagi reached for his camera, but the animal vanished. After driving back and forth along the road in search of it, however, Yagi saw it again, and this time was able to take no fewer than 19 photos from just a few feet away. He even threw a rice cake toward it, hoping to tempt it closer still, but the creature turned away and disappeared into a forest. When the pictures were examined by eminent Japanese zoologist Prof. Imaizumi Yoshinori, he felt sure that the animal was indeed a shamanu.

In November 2000, Akira Nishida took a photo of a canine mystery beast on a mountain in Kyushu that was widely acclaimed afterwards as evidence of shamanu survival—until March 2001, that is. For that was when a sign appeared on a hut on Ogata Mountain claiming that the creature in the photo was nothing more than a Shikoku-region domestic dog that the hut's owner had been forced to abandon for private reasons. However, Yoshinori Izumi, a former head of the National Science Museum's animal research section, dismisses this claim as a hoax, and remains adamant that the photographed canid is a bona fide shamanu, even stating that the pattern of the creature's fur is completely different from that of Shikoku dogs. This is evidently a saga with many more chapters still to be written.

Yet the shamanu's rediscovery would be a major cryptozoological event, so we await further news with great interest. Perhaps the howling god has not been silenced after all.

A NEW DAWN FOR THE SUN DOG?

Iconography can sometimes be a fruitful source of material for cryptozoologists. Several notable modern-day species of animal, including the gerenuk (giraffe antelope), Roxellana's snub-nosed monkey, and Grévy's zebra, were first made known to science not via the procurement of a specimen, but by depictions on ancient monuments. Nevertheless, attempting to identify a creature portrayed in

SUN DOG, A PRE-INCA PERUVIAN DEPICTION

this manner can be fraught with difficulties, as exemplified by the little-known but highly intriguing case of the Mexican sun dog.

This mysterious sacred beast is depicted in numerous paintings, carvings, woven textiles, and ceramics preserved from the bygone Mexican civilizations of the Toltecs and Aztecs, as well as the Central American Mayas, and even the Peruvian Incas. Although varying quite extensively in appearance from one site to another, the sun dog was undeniably canine in form. It generally sported a slender lengthy body, four fairly short legs whose feet were equipped with noticeable claws, a long slender tail sometimes specifically portrayed as alive, an elongate muzzle, and a long extended tongue. What might it be?

If a living specimen could somehow be obtained, its identity would soon be ascertained—or would it? In *America's Ancient Civilisations* (1953), writer Hyatt Verrill revealed that while residing in Ixtepec, southern Mexico, he was given a small animal that he felt sure was a young specimen of the Mexican sun dog. A new dawn for this ostensibly extinct enigma?

Brought to Verrill by a Lacandon Indian from Chiapas, the creature was just under two feet long, with a shortish face but lengthy slender tongue, sharp claws on its feet, short legs, a long prehensile tail, and golden-brown fur (paler ventrally). Most distinctive of all, however, was its long undershot lower jaw, jutting forth beyond the limit of its upper jaw.

The creature also possessed a facial mask, whose precise form was curiously inconsistent. Within his book, Verrill included two illustrations of his supposed sun dog pet. One was a photograph, the other was a drawing penned by Verrill himself. In the photo, the creature's mask appeared to be much paler than the fur surrounding it. Yet in Verrill's drawing, it was decidedly darker than the fur surrounding it, and resembled the black mask of a raccoon.

According to Verrill, his pet sun dog (which he nicknamed "the Monster") was very ferocious, despite its modest size and tender age, and was only too willing to attack anyone without the slightest provocation, accompanied by dramatic snarls and hisses.

I first learned of the Mexican sun dog and Verrill's pugnacious pet in September 1992, when one of my correspondents, Bob Hay from Augusta, Western Australia, sent me a copy of the relevant section from Verrill's book, containing several different ancient depictions of the sun dog, and asked my opinion regarding the living animal's identity. The ancient depictions were very interesting, and did not appear to correspond readily with any known creature alive today. As for Verrill's pet, conversely, this was, for me, a disappointingly mundane Monster, with scant resemblance to the sun dogs or any other canine beast.

Indeed, it seemed to be nothing more than a young specimen of *Potos flavus*, the kinkajou. Also called the honey bear, and measuring up to four feet long when fully grown, this golden-furred relative of the raccoons, coatis, and other procyonids has a wide distribution, extending southwards from eastern Mexico to the Mato Grosso of central Brazil—and offers a near-exact morphological match with Verrill's pet as portrayed in the photograph. Moreover, the kinkajou is the only member of the Carnivora in the whole of the New World with a prehensile tail. Not even its lesser-known lookalike relations, the olingos *Bassaricyon* spp., which are very similar to it in superficial external appearance, have a prehensile tail.

In fact, the only outward discrepancies between the kinkajou and the Monster were the latter animal's lower jaw and its savage behavior. Yet even these can be satisfactorily explained. The Monster's odd lower jaw is assuredly an abnormality—either a malformation resulting from a genetic or developmental quirk, or the product of some accident experienced by this animal, subsequently resulting in the jaw's aberrant growth. Furthermore, if such an abnormality hampered its feeding or caused pain, this could well explain the Monster's aggressive attitude. Alternatively (or additionally?), the Monster may have been treated harshly by its Indian captor, thus creating a frightened animal perpetually on the offensive.

The Monster's intimate morphological correspondence to

the kinkajou makes it all the more surprising that in his book, Verrill claimed that no zoologist had been able to identify his pet. Also, he personally ruled out any likelihood of taxonomic synonymity between this individual and either an olingo or a kinkajou (even a malformed one) because, he claimed, the Monster possessed a different (but unspecified) dentition. Yet if its lower jaw was a genetic or developmental malformation, this in turn could have affected its dentition too.

To my mind, the fundamental flaw with this entire history is Verrill's assumption that his Monster belonged to the same species as the depicted sun dogs. As far as I can see, one was nothing but a young, possibly slightly deformed kinkajou, whereas the others probably owe more to the exceedingly elaborate, highly stylized artistry of long-vanished civilizations than to any bona fide mystery beasts still eluding formal identification and classification.

ON THE SCENT OF CYNOCEPHALI

In medieval times, scholars firmly believed in the reality of all manner of extraordinary semi-human entities, supposedly inhabiting exotic lands far beyond the well-explored terrain of Europe. Prominent among these tribes of "half-men" were the cynocephali or dog-headed people, popularly referred to by travelers and chroniclers. Their domain's precise locality, conversely, depended to a large extent upon the opinion of the specific traveler or chronicler in question, as few seemed to share the same view.

According to the Roman scholar Pliny the Elder (23-79 AD), the mysterious country of Ethiopia, which he mistakenly assumed was one and the same as India, was home both to the cynocephali and to the similarly dog-headed cynamolgi. Numbering 120,000 in total, the cynamolgi wore the skins of wild animals, conversed only via canine barks and yelps, and obtained milk, of which they were very fond, not from cows but by milking female dogs.

Several centuries earlier, Herodotus (c.485-425 BC), a Greek historian, had claimed that a race of Ethiopian dog-heads termed the

Kynokephaloi not only barked instead of speaking, but also could spew forth flames of fire from their mouths!

Later authors often stated that India and nearby islands were populated by cynocephali. The 13th century Venetian traveler Marco Polo, who could never be accused of letting a good story slip by unpublicized, soberly announced that the Andaman Islands off Burma (now Myanmar) were home to a race of milk-drinking, fruit-eating (and occasionally man-devouring) dog-heads who engaged in peaceful trade with India. A century later, Friar Odoric of Pordenone relocated these entities to the nearby Nicobar Islands, whereas Friar Jordanus's writings deposited them rather unkindly in the ocean between Africa and India. Cynocephali living in India itself were reported by geographer Peter Heylyn, who traveled widely during the time of Charles I, the Interregnum, and the Restoration of Charles II.

Nor should we forget Sir John Mandeville, even though his (in)famous supposed journeys to incredible faraway lands in Africa and Asia during the 14th century owe considerably more to the imagination than to peregrination. He averred that a tribe of hound-headed people inhabited a mysterious island of undetermined location called Macumeran. Here, curiously, they venerated an ox, and were ruled by a mighty but pious king, who was identified by a huge ruby around his neck and also wore a string of 300 precious oriental pearls.

Less well known than the above-mentioned examples is the fact that dog-headed entities also feature in Celtic lore and mythology. As pointed out by Prof. David Gordon White in his definitive book, *Myths of the Dog-Man* (1991), Irish legends tell of several cynocephalic people. Perhaps the most notable of these legends concerns a great invasion of Ireland from across the western sea by a race of dog-headed marauders known as the Coinceann or Conchind, who were ultimately vanquished by the hero warrior Cúchulainn. Irish legends also claim that St. Christopher was a cynocephalus; and a tribe of dog-heads opposed King Arthur, and were fought by him (or Sir Kay, in some versions of this tale).

KING LYCAON, TRANSFORMED INTO A CYNOCEPHALUS

THE BEASTS THAT HIDE FROM MAN

Shetland folklore tells of a supernatural entity known as the wulver, which has the body of a man but is covered in short brown hair and has the head of a wolf. According to tradition, this semi-human being lives in a cave dug out of a steep mound halfway up a hill, and enjoys fishing in deep water. Despite its frightening appearance, however, the wulver is harmless if left alone, and will sometimes even leave a few fishes on the windowsill of poor folk. A similar wolf-man entity is also said to exist in Exmoor's famous Valley of the Doones.

Cynocephalic deities are not infrequent in mythology, and include such familiar examples as Anubis, the Egyptian jackal-headed god of the dead; and the Furies or Erinnyes, the terrifying trio of avenging dog-headed goddesses from Greek mythology. Originally, Anubis was the much-dreaded god of putrefaction, but in later tellings he became transformed into a guardian deity, protecting the dead against robbers, and overseeing the embalming process. His head's form was derived from the jackals scavenging in Egyptian burial graves during the far-distant age preceding the pyramids when graves were shallow and hence readily opened.

The Furies were the three hideous daughters of the Greek sky god Uranus and Gaia (Mother Earth), comprising Alecto, Megaera, and Tisiphone. Not content with canine heads, they also sported leathery bat-wings, fiery bloodshot eyes, foul breath, and hair composed of living serpents. Their allotted task was to harangue and punish evil-doers, especially parent-killers and oath-breakers, but eventually they were transformed into the kinder Eumenides.

Another figure in Greek mythology who underwent a canine transformation, though this time in reverse, was Lycaon. This wicked, foolhardy king of Thessaly served up human flesh to Zeus, in order to test whether the supreme Greek deity would recognize it. Needless to say, Zeus did, and as a punishment he changed Lycaon into a wolf. Interestingly, however, Lycaon is often portrayed not as a complete wolf, but rather as a wolf-headed man.

Zoologists seeking to nominate real animals as the inspiration for the legends of cynocephali generally offer two principal candidates. The first of these is the baboon, of which there are several species. The heads of these large monkeys are certainly very doglike; indeed, the yellow baboon is known scientifically as *Papio cynocephalus*. And the sacred baboon *P. hamadryas* is native to Ethiopia, the source of the earliest cynocephalus myths.

The second candidate is the indri *Indri indri*, the largest modern-day species of lemur, and indigenous to Madagascar. Measuring over three feet, but only possessing a very short, inconspicuous tail, and often spied sitting upright in trees, this highly distinctive creature does look remarkably like a short dog-headed human.

The indri did not formally become known to science until 1768, when French naturalist Pierre Sonnerat arrived in Madagascar. Even so, when other, earlier travelers visiting this exotic island returned home to Europe and regaled their listeners with much-embellished accounts of their journeys, these may well have included exaggerated tales about the indri.

Certainly, it would not take a sizeable stretch of the imagination to convert a dog-headed lemur into a fully fledged cynocephalus—thus breathing life into a being that never existed in reality, yet which would be faithfully chronicled by a succession of relatively uncritical scholars for many centuries thereafter. Of such, indeed, are legends all too often born.

DON'T LOSE YOUR HEAD OVER THE *WAHEELA!*

At the southern end of the Mackenzie Mountains, an area comprising coniferous forests and barren tundra, lies the Nahanni Valley, situated just inside the Canadian Northwest Territories' border with the Yukon Territory. On first sight, it may well seem a fairly nondescript, unexciting locality, but it has an evil reputation, as evinced by its nickname, "The Headless Valley." Indeed, its nickname is well earned, for over the years there have been many verified but officially unexplained cases here of prospectors and other wayfarers, found dead—and minus their heads!

To the native American Indians inhabiting this forbidding zone, however, there is no mystery. They are convinced that these deaths are the work of a monstrous wolf-like beast indigenous to northern Canada and Alaska, and referred to by them as the *waheela*.

The *waheela* first attracted widespread public attention in October 1974, when an article on this hitherto-obscure mystery beast, written by eminent cryptozoologist Ivan T. Sanderson, was published posthumously in *Pursuit*, the journal of the Society for the Investigation of The Unexplained (SITU). Sanderson revealed that at widely separate times he had received near-identical data concerning such creatures from two wholly independent colleagues.

The first to have contacted him was Tex Zeigler, a professional cameraman and film director, who had gathered information regarding the *waheela* while visiting Alaska. Around 15 years later, Sanderson had received some data describing an identical mystery beast from an experienced truck mechanic, whom he chose to identify only as "Frank" in his article. Frank claimed to have spied one of these animals while passing through the Nahanni Valley, and gave the following details concerning his close encounter of the cryptozoological kind.

Frank had left some friends at a camp below the Virginia Falls, and had journeyed up the Nahanni River with an American Indian friend whose home lay some distance further north. After a day's strenuous paddling, they made camp, and for several days they remained in this area, hunting game for food, which was plentiful here. One day, after reaching a plateau carpeted with grass and bushes but ringed by dense forests below a steep bank, the Indian decided to try and flush some game from the forest. So he told Frank to stay on the plateau, where he could readily spy and thus bag anything that emerged from the trees below.

After waiting on the plateau for a time, Frank suddenly noticed some bushes at the very edge of the forests begin to move. He called out, to ensure that it was not his companion who was responsible for this movement, but he received no response — at first. Then, without any warning, an amazing creature stealthily emerged, which Frank evocatively described as "the grand-daddy of all wolves"!

It resembled a pure-white wolf, with very long, shaggy fur, but was truly enormous, standing about three and a half feet at the shoulder. Apart from its snowy pelage, however, the most eye-catching feature of this extraordinary animal was its head, which was very wide.

After it moved to within 20 paces of him, Frank became so alarmed that he fired both barrels of his 12-gauge shotgun, loaded with birdshot and heavy ball, and was convinced that he scored a hit on the beast's left flank. However, his shot did not seem to have any effect upon the creature, which was no doubt shielded from injury by its exceedingly dense coat. Instead, it simply turned away, and leisurely ambled back into the forest. Thoroughly bemused and shaken, Frank lost no time in relating his eerie visitation to his Indian friend, but received no reply from him. It was evident that his friend knew of such animals, but considered them to be evil, ill-omened entities. Nevertheless, Frank eventually persuaded him to talk about them, and learned that they were indeed familiar to him.

The Indian stated that the creature Frank had seen was not a wolf but something completely different. Its kind are much larger than wolves and have much broader heads too, but their legs are shorter, their tails are thicker, and their ears are smaller. In addition, the toes on their feet are splayed far apart from one another, unlike those of wolves, which probably affords them a somewhat flat-footed, bear-like (plantigrade) mode of walking. Also unlike wolves, they are solitary hunters, and tend to be scavengers rather than active predators. Rarer than wolves, they spend most of the year up near the tundra region, coming south only in winter—except in the Nahanni Valley and certain other valleys to the west, where they reside throughout the year, but remain aloof from humans.

Needless to say, this description does not match that of true wolves, but it compared perfectly with the *waheela* accounts collected in Alaska by Tex Zeigler. Even more significantly, however, it also closely recalls a very distinctive taxonomic family of carnivorous mammals currently believed to have been extinct since the Ice Ages.

Known technically as amphicyonids, they are colloquially referred to as bear-dogs, and for good reason, because they seemed to combine the burly, short-limbed, plantigrade body plan of bears with the dentition and facial structure of dogs—as does the *waheela*. Moreover, bear-dogs survived in North America until at least as recently (geologically speaking) as two million years ago, and lingered on in parts of the Old World until a mere 10,000 years ago. Their ultimate extinction is believed to have been due to competition with more efficient carnivores, such as the wolf and other canids—but what would happen if a population of bear-dogs managed to survive in remote barren areas rarely ventured into by canids and human hunters, such as the tundra of northernmost North America?

Is it possible, as speculated by Sanderson, that, freed by this region's inhospitable environment from the pressures of intense interspecific competition and also persecution by humanity, a species of bear-dog has surreptitiously persisted right into the present day? Morphologically, there is no doubt that a surviving amphicyonid could provide a very persuasive identity for the mysterious *waheela*, but without a complete specimen for formal examination, we will never know for sure—and it would require an exceedingly intrepid investigator to seek out a creature that can apparently decapitate a man with a single bite!

MAKING A MONKEY OUT OF THE MIMICK DOG?

Medieval mythology is populated not only by such renowned examples of fabulous beasts as the unicorn, dragon, griffin, and basilisk, but also by a host of lesser-known yet no less fascinating fauna, which include among their number a particularly mystifying creature—the so-called "mimick dog."

It was said to originate in the Middle East, particularly in the Libyan province of Getulia, and a concise account of this peculiar animal can be found in Edward Topsell's bestiary, *The Historie of Foure-Footed Beastes* (1607). According to Topsell, the mimick dog is:

> ...apt to imitate al things it seeth, for which cause
> some have thoght, that it was conceived by an
> Ape, for in wit and disposition it resembleth an
> ape, but in face sharpe and blacke like a Hedghog,
> having a short recurved body, very long legs,
> shaggy haire, and a short taile: this is called of
> some *Canis Lucernarius*. These being brought up
> with apes in their youth, learne very admirable
> and strange feats, whereof there were great plenty
> in *Egypt* in the time of king *Ptolemy*, which were
> taught to leap, play, and dance, at the hearing of
> musicke, and in many poore mens houses they
> served insteed of servants for divers uses.

Any dog whose muddled morphology combined the face of a hedgehog with the body of an ape would be a wonder to behold, but its talent for mimicry set the mimick dog even further apart from the typical canine creed. Consequently, they were much sought after by traveling players and puppeteers, who would train them to participate in their performances.

One such performance took place during a public spectacle at Rome, attended by the emperor Vespasian. It featured an extremely versatile mimick dog that effortlessly imitated the behavior and cries of a diverse range of different types of dog and other animals too. Its pièce de resistance, however, was its own tragic death, after eating some poisoned bread—and its enthusiastic resurrection from the dead at the end of the play, which it performed with great verve, delighting the emperor and the other members of the audience, who included the famous Greek biographer Plutarch from the first century AD.

Yet in spite of its erstwhile popularity, today the mimick dog is largely forgotten, and even among those few zoologists aware of its history as recorded in ancient and medieval works there is no firm agreement concerning the precise identity of this extraordinary little animal. Indeed, there is no guarantee that it was actually a dog at all.

MIMICK DOG

In *Curious Creatures in Zoology* (1890), John Ashton believed that it was a poodle, but the picture reproduced here, from Topsell's bestiary, is hardly reminiscent of this familiar breed, or even of some ancestral form. And it would be a very clever poodle indeed that could adequately perform the tasks normally carried out by human servants. Similarly, for all sorts of fundamental genetic reasons, the suggestion made by some early authorities that mimick dogs were the product of illicit liaisons between dogs and apes does not merit even the briefest of considerations.

A far more likely explanation for this furry caricaturist is that the mimick dog was not of the canine persuasion at all. Certainly, its gift for accurate impersonation and mimicry readily recalls a monkey. More specifically, its ape-like body, long limbs, dense fur, and slender muzzle are all consistent with baboons. One species of baboon occurs in Egypt (though not in Libya), from where specimens accompanying traveling performers could have reached Europe (including England, where mimick dogs were indeed present, according to Topsell).

Such an identity may even explain the strange idea that the mimick dog was a simian-canine crossbreed, because baboons have monkey-like bodies but very dog-like heads. As already alluded to when discussing the cynocephali, exemplifying this latter characteristic are the yellow baboon, referred to scientifically as *Papio cynocephalus*, and the anubis baboon, named after Anubis, ancient Egypt's jackal-headed god of embalmers and death.

Baboons are easily trained to perform tricks, they are often kept as intelligent pets, and sometimes have even been taught to carry out tasks normally reserved for humans. Perhaps the most famous example was a baboon from the late 19th century called Jack, who worked for most of his life as a signalman, assisting his crippled human master to operate the railway line signals at a small train station in South Africa. Throughout his amazing career, which spanned nine years, Jack never made a single mistake! During the 1970s, another baboon was fulfilling the same role at a railway station near Pretoria, and there was a baboon elsewhere in South Africa whose farmer owner had trained him as a shepherd!

Nevertheless, the creature depicted in Topsell's book bears no resemblance to a baboon—even allowing for the fact that medieval illustrations of animals were often notoriously inaccurate. Nor does it look like the Barbary ape *Macaca sylvanus*, a tailless species of macaque monkey native to North Africa (including Libya long ago) but renowned for the largely human-introduced colony on Gibraltar. In fact, Topsell's picture does not recall any type of monkey.

At present, therefore, verification of the mimick dog as a baboon remains unproven, but perhaps this is fitting. After all, it would be nothing if not ironic if science quite literally made a monkey out of a creature best known for aping around.

CHAPTER 14:

Lemurs of the Lost— and Found?

In Madagascar fortunate circumstances have almost enabled us to watch the extinction of the giant fauna of the past, but we have missed our opportunity.

BERNARD HEUVELMANS—*ON THE TRACK OF UNKNOWN ANIMALS*

BUT HAVE WE REALLY MISSED OUR OPPORTUNITY? Since the early 1980s, several species of lemur have been newly discovered, or rediscovered after being presumed extinct, on their island homeland, Madagascar. Moreover, there could be some dramatic cryptozoological encounters of the lost lemur kind still to be made here.

FLACOURT'S *TRATRATRATRA*

Take, for instance, the *tratratratra*. This mysterious beast was first brought to Western attention by Admiral Étienne de Flacourt in his *Histoire de la Grande Isle Madagascar* (1658). He described it as a very solitary animal: "as big as a two-year-old calf, with a round head and a man's face; the forefeet are like an ape's, and so are the hindfeet. It has frizzy hair, a short tail and ears like a man's." He also noted that a *tratratratra* had been seen near the Lipomani lagoon, and revealed that the native people and this sizeable creature shared a great fear of one another, each fleeing when catching sight of the other.

No present-day Madagascan mammal fits the *tratratratra's* description, but it does sound very reminiscent of a supposedly long-extinct giant lemur known as *Palaeopropithecus*, which was as big as a chimpanzee and probably spent at least some of its time on the ground because of its large size. In particular, its flattened face would have looked ape-like or even humanoid, thus differing noticeably from the arboreal, dog-faced lemurs predominating in Madagascar today, but readily recalling the *tratratratra*. *Palaeopropithecus* supposedly became extinct one thousand years ago, due to hunting and habitat destruction, but some zoologists believe that it may have persisted a few centuries longer, explaining reports of the *tratratratra*.

Indeed, it is not impossible that this extraordinary "lost" lemur has actually lingered into the 20th century. As recently as the 1930s, a French forester in Madagascar claimed to have observed at close range an amazing, unidentified lemur unlike any that he had ever seen before. According to his description, it resembled a gorilla, sat four feet high, lacked a muzzle, and possessed "the face of one of my ancestors"! Was it a surviving *Palaeopropithecus*?

TOKANDIA—TITAN OF THE TREES

Equally controversial is the *tokandia*, which is said to be a huge quadrupedal jumping beast that climbs up into trees and cries like a man, but does not have a man-like face. French cryptozoologist Dr. Bernard Heuvelmans has likened this remarkable mystery beast to a real-life Madagascan monster of sorts, *Megaladapis*, the largest lemur of all time.

Probably weighing over 200 pounds, *Megaladapis* was the size of a small bear, and its skull was rather like that of a rhinoceros but with heavy cow-like jaws. It may have sported a prehensile tapir-like snout too, for pulling tasty foliage towards its mouth. Yet despite its great size, it is believed to have been arboreal, clinging to the trees like a colossal koala.

So could *Megaladapis* and the *tokandia* be one and the same? Like *Palaeopropithecus*, *Megaladapis* "officially" died out at least a millennium ago, but surely it is more than a coincidence how closely this "lost" lemur compares with the tantalizing *tokandia*?

KIDOKY—YET ANOTHER LIVING FOSSIL?

No less fascinating is the *kidoky*. December 1998 saw the publication in the journal *American Anthropologist* of a remarkable scientific paper, by Fordham University biologist Dr. David A. Burney and Madagascan archaeologist Ramilisonina, in which they documented ethnographic data collected by them during late July-early August 1995 at the remote southwestern Madagascan coastal fishing village of Belo-sur-mer. Their paper contained accounts by several local eyewitnesses describing what they claimed were quite recent sightings of two scientifically unidentified creatures. These are known respectively as the *kilopilopitsofy* ("floppy ears"—also termed the *tsomgomby*—which may be a relict representative of one of Madagascar's two officially extinct species of pygmy hippopotamus) and the *kidoky*. Witnesses were questioned independently of each other, and prompting for specific answers was avoided wherever possible. A measure of how zoologically accurate and trustworthy the knowledge and testimony of each witness were likely to be was ascertained by asking each witness to identify local faunal species from unlabeled color plates. The results obtained generally satisfied the authors that the eyewitnesses were indeed knowledgeable and trustworthy concerning their region's fauna.

Of the seven major interviewees whose details were tabulated in the paper, the most authoritative and informative was an old villager called Jean Noelson Pascou, who had keen eyesight and was literate. When questioned about the *kidoky*, Pascou, and also some recent woodcutter eyewitnesses, compared it in basic form with those moderately large, primarily arboreal lemurs known as sifakas, but the *kidoky* was said to be much bigger. Based upon a close sighting of one in 1952, Pascou claimed that it has dark fur but a white spot on its brow and another below its mouth. Apparently, the *kidoky* is usually encountered on the ground, and flees by running (via a series of short, forward, baboon-like leaps, instead of via the characteristic sideways leaping gait always adopted by sifakas when on the ground), rather than by taking to the trees. Its long whooping call is like that of the large dog-headed indri, but it has a rounder man-like face, more like a sifaka's than an indri's.

As conceded by Burney and Ramilisonina, this description brings to mind two officially extinct genera of giant lemur, *Archaeolemur* and *Hadropithecus*. Both possessed many terrestrial modifications, and have been likened to baboons anatomically and behaviorally. Conversely, radiocarbon datings place both as having been extinct for well over 300 years at least, probably longer —but have they?

SIFAKA

HABÉBY AND *MANGARSAHOC*—LEMURS IN SHEEP'S CLOTHING?

Also in need of an explanation are the *habéby* or *fotsiaondré* reported from the Isalo range by the Betsileo tribe, and the *mangarsahoc* of Anosy. The *habéby* is said to be a nocturnal sheep with long furry ears, a white coat dappled with black or brown, a long muzzle, and large staring eyes. Yet no one has ever reported seeing a horned *habéby*, i.e. a *habéby* ram. In view of this odd omission, and also bearing in mind that whereas sheep are not normally nocturnal and do not have large staring eyes, while many lemurs are and do, zoologists consider it much more likely that the *habéby* was (or is?) an unknown species of large lemur.

This identity can also be applied to the *mangarsahoc*, traditionally described as a very large nocturnal ass or horse but with such enormous ears that they hang down over its face. As with the *habéby*, it is said to bring bad luck to anyone who sees it—and similar superstitions are attached to many of the lemur species known today.

RESURRECTING THE GIANT AYE-AYE?

Even the bizarre-looking aye-aye *Daubentonia madagascariensis*, resembling a goblinesque squirrel, may have a cryptozoological counterpart. There was once a larger version, the giant aye-aye *D. robusta*—whose skeletal remains indicate that it was a third bigger than the typical aye-aye. Although the giant aye-aye supposedly died out many centuries ago, there is an intriguing piece of evidence to suggest that this noteworthy species may have still been alive much more recently. A scientific paper from 1934 noted that near the village of Andranomavo in the Soalala District, a government official called Hourcq saw a native hunter holding the skin of an exceptionally large aye-aye—so large, in fact, that some researchers feel that it might indeed have been from a modern-day specimen of the giant aye-aye.

"Lemur" is a Malagasy word meaning "ghost," but judging from the reports reviewed here, some of Madagascar's "lost" lemurs may be far more substantial than any ghost!

CHAPTER 15

"Bring Me the Head of the Sea Serpent!"

SEA SERPENT: *No one has counted his genials and gastroliges yet, nor computed his parietals and described his superaoculars, though there's plenty would like to and pickle him to boot and stick him in a museum for people to peer at.*

CHARLES G. FINNEY—*THE CIRCUS OF DR. LAO*

OFTEN UTTERED WITH NO LESS VEHEMENCE THAN Salome's demand for the head of John the Baptist, the title of this chapter reflects the response traditionally elicited from most zoologists when confronted with the suggestion that the vast oceans may contain notably large species of sea creature still unknown to science—and that some such species could explain that most controversial maritime enigma, "the great sea serpent."

Needless to say, fleeting glimpses of a head and neck (or sometimes merely an ill-defined series of humps) all-too-briefly surfacing before sinking back beneath the waves can hardly substitute for adequate *physical* evidence—a head, or a body!—that can be subjected to formal scientific analysis in order to determine categorically the taxonomic identity of its owner. One might assume, therefore, that if such evidence does come to hand, the end of the mystery must surely then be in sight. But as readily demonstrated by the following selection of cases, the course of true cryptozoology seldom runs quite as smoothly as that!

STRONSAY'S SIX-LEGGED SEA SERPENT

Possibly the most famous of these cases occurred at Stronsay, one of the Orkney Islands off northern Scotland. On September 26, 1808, John Peace was fishing east of Rothiesholm Point when he saw what seemed to be the carcass of a whale, cast up onto the rocks, above which were flocks of circling seabirds. He rowed up to it in his boat and examined it, and found that it was a very peculiar-looking creature which did not resemble anything known to him. At that same time, farmer George Sherar was watching Peace from the shore, and was able to confirm all of this. About 10 days later, moreover, he was able to see it for himself, because it was washed ashore on Stronsay, lying on its belly just below the high tide mark.

When Sherar discovered it there, he measured it, and found it to be 55 feet long. Others also measured it, and obtained the same result. It was very serpentine, almost eel-like in general build, but possessed a 15-foot neck, a small head, and a long mane running along its back to the end of its tail. Most bizarre of all, however, was that it seemed to have three pairs of legs, and each foot had five or six toes. Sherar salvaged some vertebrae and the skull of this extraordinary creature, which became known as the Stronsay beast.

Details of its discovery and description ultimately reached Patrick Neill, secretary of Edinburgh's Wernerian Natural History Society, and at a meeting of the society on November 19, 1808, Neill released some details on this subject. At the next meeting, on January 14, 1809, he gave the Stronsay beast a formal scientific name: *Halsydrus pontoppidani*, "Pontoppidan's water snake of the sea," after Erik Pontoppidan, an 18th century Norwegian bishop who had collected many sea serpent reports.

At that same meeting, a Dr. John Barclay, who had examined some of the beast's remains in Orkney, presented a paper in which he described the vertebrae, skull, and one of the creature's legs. His paper, accompanied with detailed diagrams, was published in 1811, within the society's memoirs, and attracted a great deal of attention.

The vertebrae were very striking, resembling cotton reels, and were cartilaginous, but with calcification that radiated from the center of each vertebra in a star-like pattern. The leg was also cartilaginous, but was not a real, jointed leg at all; it was merely a fin.

To many people, these features meant little, but they meant a great deal to the eminent naturalist Sir Everard Home, who was working at that time upon an exhaustive study of the basking shark *Cetorhinus maximus*, the world's second-largest species of shark, but generally harmless, living on plankton. When Home heard about the Stronsay beast, he felt sure that it must have been a shark. This is because the only creatures to have cartilaginous vertebrae are sharks and rays, and the only creatures to have cartilaginous vertebrae with star-shaped calcification are sharks. Furthermore, when he compared the Stronsay beast's vertebrae and other salvaged remains with the corresponding portions of a known specimen of basking shark, they matched very closely.

But the long-necked, six-legged, mane-bearing Stronsay beast *looked* nothing like a basking shark—so how could this drastic difference in appearance be resolved? In fact, it was quite simple. When basking shark carcasses begin to decompose, the entire gill apparatus falls away, taking with it the shark's characteristic jaws, and leaving behind only its small cranium and its exposed backbone, which have the appearance of a small head and a long neck. The triangular dorsal fin also rots away, sometimes leaving behind the rays, which can look a little like a mane—especially when the fish's skin also decays, allowing the underlying muscle fibres and connective tissue to break up into hair-like growth. Additionally, the end of the backbone only runs into the top fluke of the tail, which means that during decomposition the lower tail fluke falls off, leaving behind what looks like a long slender tail. The pectoral and sometimes the pelvic fins remain attached, but become distorted, so that they can (with a little imagination!) look like legs with feet and toes, and male sharks have a pair of leg-like copulatory organs called claspers, which would yield a third pair of legs. Suddenly, the basking shark has become a hairy six-legged sea serpent.

Over the years, almost all of the Stronsay beast's preserved remains have been lost or destroyed, but three vertebrae are retained in the Royal Museum of Scotland, the last remnants of the world-famous Stronsay sea serpent. Its mystery, conversely, continues to the present day, and for very good reason. The longest *conclusively* identified basking shark that has been accurately measured was a truly exceptional specimen caught in 1851 in Canada's Bay of Fundy; whereas the average length for its species is under 26 feet, this veritable monster was a mighty 40 feet, three inches. Yet even that is almost 15 feet less than the Stronsay beast. Even the largest scientifically measured specimen of the world's biggest fish—the whale shark *Rhincodon typus*—was only(!) 41.5 feet long. So was the 55-foot Stronsay beast really a basking shark, or could it have belonged to a still-unknown, giant relative?

Bearing in mind that one of the world's largest known sharks, the formidable megamouth *Megachasma pelagios*, remained wholly unknown to mankind until November 15, 1976, when the first recorded specimen was accidentally hauled up from the sea near the Hawaiian island of Oahu, the prospect of undiscovered species of giant sharks still eluding scientific discovery is far from being as unlikely as one might otherwise assume.

In December 1941, history repeated itself in the Orkneys when a strange carcass, 25 feet long, was washed ashore at Scapa Flow. Its superficially prehistoric appearance was presumably sufficient for Provost J.G. Marwick, who had documented it in detail in a local newspaper account (*Orkney Blast*, January 30, 1942), to dub this enigma "Scapasaurus." Fortunately, a single vertebra from its remains was preserved and retained at the British Museum (Natural History), which identified it as a basking shark.

PSEUDOPLESIOSAUR OF THE *ZUIYO MARU*

On April 25, 1977, the decomposing carcass of a huge, plesiosaur-like beast was caught in the nets of the Japanese fishing vessel *Zuiyo*

Maru, trawling in waters about 30 miles east of Christchurch, New Zealand. The carcass, about one month dead, measured roughly 33 feet long, was in an extremely advanced state of decomposition, and smelled so awful that the crew very speedily cast it overboard, but fortunately not before its measurements were recorded, a few fibres were taken from its carcass, and some photos were obtained of it.

Biologists were very perplexed by the photos, and all manner of identities were offered, ranging from a giant sea lion, or a huge marine turtle, to the popular plesiosaur identity, or a whale. For quite a time, its true identity remained undisclosed, until the fibres from it were meticulously analyzed by Tokyo University biochemist Dr. Shigeru Kimora. During his research, he discovered that the samples contained a very special type of protein, known as elastoidin, which is only found in sharks and rays. Once again, the scientific world had been fooled by a deceptively shaped decomposing shark carcass—equally, however, it was a carcass of remarkable length. Just *another* exceptional basking shark?

THE HORNED SEA SERPENT OF GREATSTONE

On April 14, 1998, two beachcombing boys, Peter Jennings and Neil Savage, discovered a strange, unidentified, decomposing carcass washed ashore at Greatstone in Kent, England. Measuring about eight feet long, it still possessed a skull, a series of large vertebrae, and some rotting body tissue. Intriguingly, however, the skull was said to bear a pair of short curved horns! Tragically, council workmen disposed of the carcass before tissue or skeletal samples could be taken from it and examined, but some good photographs were obtained. Not long afterwards, Fortean writer Paul Harris contacted me concerning this putative sea serpent, kindly supplying me with a couple of newspaper cuttings (*Kentish Express*, April 19, 1998; *Folkestone Herald*, April 23, 1998) that contained photos of its remains, plus various additional details that he had gathered during his own investigation of this case.

As soon as I saw the photos, I realized that the carcass was strikingly similar to the famous Hendaye sea serpent corpse of 1951, which had been conclusively identified by cryptozoologist Dr. Bernard Heuvelmans in his classic tome *In the Wake of the Sea-Serpents* (1968) as that of a basking shark. All of the telltale features were present—the cottonreel-shaped vertebrae, the long triangular snout, and, most distinctive of all, a pair of slender curling "horns" or "antennae" projecting from the snout's base. These are in fact the rostral cartilages that, in life, raise up the shark's snout. However, one can readily appreciate that to eyewitnesses not acquainted with shark anatomy (particularly *decomposing* shark anatomy!), the presence of "horns" might well be enough to make them assume that a strange-looking carcass like this one must surely be something more exotic than a rotting basking shark.

SCOLIOPHIS ATLANTICUS

Another potentially promising sea serpent corpse that suffered a humiliating demotion in status was the three-foot-long "baby sea serpent" captured in autumn 1817 in a field near the harbor of Cape Ann, north of Gloucester, Massachusetts, which had recently been the scene of numerous multiple-eyewitness sightings of an enormous snake-like sea monster that sometimes appeared humped. When the "baby" specimen was captured, it was deemed by a committee of scientists from the Linnaean Society of New England to be a young specimen of the monster that had lately visited Cape Ann, and was duly christened *Scoliophis atlanticus* ("Atlantic humped snake"). When dissected and examined by fish researcher Alexandre Lesueur, however, it was revealed to be nothing more dramatic than a deformed specimen of a common American snake called the black racer *Coluber constrictor*.

A VERTICAL CHALLENGE FOR SEA SERPENTS?

One of the most popular "orthodox" zoological identities for sea serpents, especially some of the elongate, serpentiform versions, is the oarfish *Regalecus glesne*, and a number of snake-like carcasses washed ashore over the years have indeed proven to be specimens of this extraordinary creature. Its candidature as an identity for living sea serpents, conversely, has lately diminished thanks to a chance but quite remarkable discovery.

The world's longest species of bony fish, the oarfish has traditionally been thought to swim horizontally, propelled via sinuous lateral undulations. As he revealed in an illustrated account published by the English magazine *BBC Wildlife* (June 1997), however, during a recent dive off Nassau, in the Bahamas, Brian Skerry was fortunate enough not only to encounter a living oarfish at close range, but also to photograph it. He was amazed to discover that it did not swim horizontally at all. Instead, it held its long thin body totally upright and perfectly rigid, with its pelvic rays splayed out to its sides to yield a cruciform outline, and it seemed to propel itself entirely via movements of its dorsal fin.

Oarfishes may swim horizontally too, but until now no one had ever suspected that this serpentine species could orient itself and move through the water in this odd, perpendicular fashion—a mode wholly distinct from the versions reported for elongate sea serpents.

THE MONSTER OF MASBATE

Staying with elongate sea serpents: in February 1997, the central Philippine province of Masbate was the focus of intense cryptozoological interest, following the news that the carcass of a mysterious sea monster had been washed ashore here on Christmas Day 1996, in the coastal town of Claveria. According to press reports (e.g. *South China Morning Post,* February 25, 1997), the carcass was of a 26-foot-long eel-like creature with a turtle-like head. A photograph of its preserved skull, vertebrae, and limbs appeared in the *Philippine Star.*

This photo, together with portions of its dried flesh, was presented to various unnamed experts for examination, but no one could identify them. According to Philippines University zoologist Dr. Perry Ong: "Judging from what I see now, it's an eel-like fish.

It must be an ancestral or primitive fish. It had fins. But if it's a fish, where are the ribs? It is not a mammal." Curiously, however, its head apparently possessed a blowhole-like orifice, resembling that of a dolphin, though it lacked the latter's familiar elongate beak (rostrum). My own opinion is that it was probably a highly decomposed killer whale *Orcinus orca*.

Perhaps the most mystifying aspect of all concerned the alleged suggestion by some "scientists" that its bones should be carbon-dated: "to determine if the animal is prehistoric." Yet as the creature evidently died only very recently, as evinced, for instance, by the presence of dried flesh still attached to its carcass, it is clearly not prehistoric. So why on Earth would any scientist propose carbon-dating it? Since this brief burst of media interest, no further news has emerged regarding the Masbate monster or the fate of its carcass.

GOUROCK SEA SERPENT

The history of cryptozoology is embarrassingly well supplied with classic cases of lost opportunities, and the long-running saga of the sea serpents has provided quite a number of these over the years. One of the most notable examples took place on the shores of Gourock, on Scotland's River Clyde. This was where, in summer 1942, an intriguing (if odiferous!) carcass was stranded that was closely observed by council officer Charles Rankin.

LINE-DRAWING OF THE GOUROCK SEA SERPENT BASED UPON
AN EYEWITNESS'S SKETCH

Measuring 27 to 28 feet long, it had a lengthy neck, a relatively small flattened head with a sharp muzzle and prominent eyebrow ridges, large pointed teeth in each jaw, rather large laterally sited eyes, a long rectangular tail that seemed to have been vertical in life, and two pairs of L-shaped flippers (of which the front pair were the larger, and the back pair the broader). Curiously, its body did not appear to contain any bones other than its spinal column, but its smooth skin bore many six-inch-long bristle-like "hairs" resembling steel knitting needles in form and thickness but more flexible.

Rankin was naturally very curious to learn what this strange creature could be, whose remains resembled those of a huge lizard in his opinion. But as World War II was well underway and this locality had been classed as a restricted area, he was not permitted to take any photographs of it, and scientists who might otherwise have shown an interest were presumably occupied with wartime work. Consequently, this mystery beast's carcass was summarily disposed of—hacked into pieces, and buried in grounds that have since been converted into a football pitch. All that remained to verify its existence was one of its strange "knitting needle" bristles, which Rankin had pulled out of a flipper and kept in his desk, where it eventually shriveled until it resembled a coiled spring.

When considered collectively, features such these bristles (readily recalling the ceratotrichia—cartilaginous fibres—of shark fin rays), the lizard-like shape, vertical tail (characteristic of fish), lack of body bones, and smooth skin suggest a decomposing shark as a plausible identity. Yet the large pointed teeth argue against the traditional basking shark explanation in favor of one of the large carnivorous species. However, if Rankin's estimate of its size was accurate, it must have been a veritable monster of a specimen—the world's largest known species of carnivorous shark, the notorious great white shark *Carcharodon carcharias*, rarely exceeds 20 feet. If only some taxonomically significant portion of the Gourock sea serpent's body could have been retained for formal examination, in particular its skull, a flipper, or at least some teeth. Instead, they have presumably been pounded ever deeper into the earth by the studs of a succession of soccer teams—oblivious to the cryptozoological treasure lying forgotten beneath their feet.

THE HEAD OF A SEA SERPENT AT LAST?

And finally: it would be quite unthinkable to end this chapter without discussing the highly controversial case of the Monongahela monster. For if the case is genuine (not a hoax, as some authors have suggested), one ship successfully obeyed the imperious command of this chapter's title—by obtaining for scientific scrutiny the head of a sea serpent! On January 13, 1852, while in latitude 3°10'S and longitude 131°50'W, the whale ship *Monongahela* of New Bedford encountered an enormous serpentine creature longer than the 100-foot ship itself, and just under 50 feet in diameter, with a 10-foot-long alligator-like head whose jaws contained 94 teeth (each approximately three inches long and recurved like a snake's).

During a titanic struggle, the ship's sailors sought to capture their monstrous visitor by harpooning it; the next morning its lifeless carcass, brownish-yellow and 103 feet seven inches long, rose to the surface of the sea. Although giant snakes are not believed nowadays to be responsible for any of the various types of sea serpent reported over the years, this particular specimen did possess some distinctly ophidian characteristics, including its recurved teeth, a lower jaw whose bones were separate, and two lungs of which one was notably larger than the other. However, it also exhibited some highly *un*-snake-like features, such as a pair of whale-like blow-holes, and four paw-like projections of hard, loose flesh.

Taxonomic considerations notwithstanding, it was clearly impractical to attempt to preserve the gigantic creature's entire carcass, so the sailors hacked off its ferocious-looking head for retention as absolute proof of this astonishing beast's reality. The *Monongahela's* master, Captain Charles Seabury, prepared a detailed account of the whole incident, including a full description of the creature itself; on February 6, the *Monongahela* encountered the brig

Gipsy, journeying to Bridgeport, so Seabury handed his account to the *Gipsy's* master, Captain Sturges, who promised to hand it into Bridgeport's post office when the *Gipsy* arrived there. Presumably he kept his word, because various newspaper accounts of Seabury's report duly appeared, including one in London's *Times* for March 10, 1852.

And this is where, for over a century, the story ended—because nothing more was heard of either the sea serpent head or the *Monongahela* carrying it. Accordingly, some cryptozoologists discounted the whole affair as an elaborate hoax, until 1959, which saw the publication of Frank Edward's book *Stranger Than Science*. This revealed that the ship carrying back Seabury's account had actually been the *Rebecca Sims*, with a Captain Gavitt as its master, and that Seabury's Christian name was Jason, not Charles. In addition, Edwards had learned that many years after Seabury's account had hit the headlines, the name board of the *Monongahela* had been discovered on the shore of Umnak Island in the Aleutians. So what had happened to the ship? As no other trace of it has apparently been found, if the incident was indeed genuine did some catastrophe occur during its continuing voyage that consigned the *Monongahela* and its entire crew to the bottom of the sea, thereby returning its unique cryptozoological cargo from whence it had come, the unknown ocean depths?

As with so many other cases on record within the ever-increasing chronicles of the sea serpent, the chances are that we will simply never know.

CHAPTER 16

A Supplement To Dr. Bernard Heuvelmans's Checklist of Cryptozoological Animals

The truth is, that folk's fancy that such and such things cannot be, simply because they have not seen them, is worth no more than a savage's fancy that there cannot be such a thing as a locomotive, because he never saw one running wild in the forest.

CHARLES KINGSLEY—*THE WATER-BABIES*

THE FOLLOWING CHAPTER CONSTITUTES IN ITS entirety and in largely unmodified form the cryptozoological checklist supplement of mine that was originally published in 1998 within *Fortean Studies* (vol. 5), and includes its accompanying self-contained bibliography.

Summarizing more than a hundred cryptids, Dr. Bernard Heuvelmans's "Annotated Checklist of Apparently Unknown Animals With Which Cryptozoology Is Concerned" (Heuvelmans, 1986) was arguably his most significant publication since *Sur la Piste des Bêtes Ignorées* (Heuvelmans, 1955). A slightly updated version, in French, was published in 1996, containing a few amendments (Heuvelmans, 1996). Otherwise, more than a decade has passed since this list originally appeared, during which period a considerable number of additional cryptids, previously unknown even to cryptozoologists, have attracted attention.

Some of these are mystery animals that had been documented in publications only recently perused for the first time by readers knowledgeable in cryptozoological matters; hence cryptozoologists had not been aware of their existence before. In other instances, mystery beasts never recorded in any publication, cryptozoological or otherwise, have been mentioned during conversations between local people and visiting scientists, explorers, or travelers.

As someone who has been working on a full-time professional basis in cryptozoology for many years, I have amassed an archive of material dealing with these hitherto overlooked cryptids that is now sufficiently sizeable to warrant the preparation of the first detailed supplement to Heuvelmans's checklist, as now presented here. To facilitate consistency, the format of this supplement adheres to that of the original checklist wherever possible.

Similarly, necessarily echoing portions of Heuvelmans's own introduction to his list, I must point out that this supplement is by no means complete. Obviously, I have not been able to read through every single book and journal ever published in search of cryptozoologically relevant reports. Equally, however, I have refrained from including examples that I do have on file but which I consider to be dubious or too vague to warrant coverage here. Also, please bear in mind that the inclusion in this supplement of a given cryptid does not guarantee its distinction from all living taxa currently recognized by science. Some may merely be morphs or aberrant individuals of recognised taxa. Furthermore, however rigorously any cryptozoological case is examined, the shadow of possible misidentification or outright fraud can never be entirely dispelled in the absence of conclusive physical evidence to examine.

Unlike Heuvelmans, I have included various creatures that officially became extinct during the earlier part of the 20th century but which may still survive today. In such cases, their continuing existence has yet to be recognized by science but they are known to local people, thereby complying with Heuvelmans's own definition of a cryptid (Heuvelmans, 1982; Greenwell, 1984). In addition, they include examples that have been extensively investigated and hence have focused public attention on cryptozoological research, such as the thylacine (Tasmanian wolf) and the eastern cougar.

Amendments to some of Heuvelmans's proposed identities for certain cryptids in his checklist have been proposed and published by fellow cryptozoologists since 1986. The most significant of these are presented here in an addendum, together with additional amendments and alternative identities that I consider merit mentioning in any supplement to that list.

Lastly: in contrast to Heuvelmans's checklist, which did not contain bibliographical references for all of the cryptids listed in it, every cryptid account included here is accompanied by references that can be consulted for further details.

MARINE FORMS

Extraordinary marine cryptids characterized by a profuse covering of white fur, a lengthy tail, and a long elephantine trunk or trunk-like proboscis at the anterior end of their body. The carcass of one such creature was washed ashore at Margate, South Africa, in November 1922; another was beached on Alaska's Glacier Island in November 1936 (Heuvelmans, 1968; Chorvinsky, 1995; Shuker, 1997d). It could be argued that these were merely decomposing sharks or whales, and that their white fur was nothing more than dried-out connective tissue. If this is true, however, the manner in which decomposition can "create" their anomalous trunk-like appendage remains unresolved. In any event, a sizeable number of eyewitnesses had observed the Margate specimen while still alive, battling with two whales—and fully furred. Only later did its dead body wash ashore. Clearly, therefore, its fur cannot be explained away as decomposing connective tissue. Based upon eyewitness descriptions, the morphology of these trunked sea monsters cannot be reconciled with any known species, living or fossil. They are presumably mammalian, but without physical remains to examine it is not possible to progress further at this time in search of a satisfactory identity.

Unidentified mysticete (baleen) whale distinguished from all known species by possessing two dorsal fins, both erect, triangular, and well developed. Roughly 60 feet long, with greenish-grey back, long falciform flippers, and greyish-white underparts. Dubbed the rhinoceros whale, its principal record is a sighting made off the coast of Chile on September 4, 1867, by Italian naturalist Prof. Enrico Giglioli, who named this species *Amphiptera pacifica*. Reports of similar beasts have since been logged off Scotland, and from the Mediterranean near Corsica (Raynal & Sylvestre, 1991; Heuvelmans, 1996; Shuker, 1997d).

Mysterious greyish-black whale, roughly 25 feet long, and sporting a very distinctive, erect dorsal fin, at least five feet high, on the highest part of its back, sighted near Chilaw on Sri Lanka's western coast on April 7, 1868, by E.W.H. Holdsworth, a Fellow of the Zoological Society of London. He later learned that this unidentified whale was familiar to his vessel's crew, who called it the "Palmyra fish" (its fin presumably reminding them of the tall erect palmyra tree common in their native Malabar home). Holdsworth was convinced that it was not a killer whale, the only known cetacean with a dorsal fin resembling his whale's (Holdsworth, 1872; Shuker, 1997d).

Unidentified dolphin with brown upperparts and white belly, two specimens spied by Sir Peter Scott on February 4, 1968, among a school of piebald dolphins *Cephalorhynchus commersonii* in the Magellan Straits. They could simply have been white-bellied dolphins (formerly *C. albiventris*); if not, they may represent a new species (Scott, 1983; Shuker, 1997d).

Single-tusked, narwhal-like cetacean spied in the Antarctic Ocean's Bransfield Strait on December 17, 1892, by crewmen aboard the *Balaena* during the Dundee Antarctic Expedition of 1892-1893. As *Monodon monoceros*, the only known living species of narwhal, is confined to Arctic waters, if their sighting was accurate there would seem to be an undiscovered southern counterpart in the Antarctic Ocean (Murdoch, 1894; Shuker, 1997d).

Dimorphic mystery ziphiid (beaked whale) often reported from the eastern tropical Pacific, occurring as a black-and-white morph and as a uniformly grey-brown morph. Dr. Robert L. Pitman has suggested that if not a new species or subspecies, it may be Longman's beaked whale *Indopacetus pacificus* (Pitman, 1987; Shuker, 1993b).

Alleged manatee-like aquatic mammals reported from sea around the mid-Atlantic island of St. Helena. It has been suggested that these are elephant seals, not sirenians, but no formal investigation has so far been undertaken (Lydekker, 1899; Shuker, 1993a).

Possible existence of unrecorded populations of the coelacanth *Latimeria chalumnae*, or even an undescribed related species, in the Mexican Gulf, and perhaps also in the coastal waters of Easter Island (Fricke, 1989; Shuker, 1993b; Raynal & Mangiacopra, 1995; Shuker, 1995f).

Giant rat-tails (macrourids)—one, six feet long, spied in deepwater near Bermuda in 1930s; another, almost 10 feet long, just above Mexican Gulf seafloor in late 1960s (Soule, 1981).

Several named but still uncollected, unphotographed species of fish spied and documented by biologist William Beebe while observing marine fauna 2,100 feet beneath the surface of the sea in a bathysphere located five miles southeast of Bermuda's Nonsuch Island. These include *Bathysphaera intacta* (a highly novel species of melanostomiatid), *Bathyceratias trilynchus* (angler fish with three "fishing rods"), *Bathyembrix istiophasma* (bizarre torpedo-shaped fish with large triangular posterior fins), *Bathysidus pentagrammus* (five-starred butterfly fish), and the scientifically unnamed abyssal rainbow gar (Beebe, 1934; Shuker, 1993b).

An unusual but very large form of shark, lacking a prominent dorsal fin, and lying in wait for its prey on the sea floor; known locally as the ground shark, and inhabiting the Timor Sea. Conceivably a giant form of wobbegong or carpet shark (Ley, 1941; Shuker, 1997d).

Colossal sharks occasionally spied in the South Pacific, closely resembling the great white shark *Carcharodon carcharias* but much longer and sometimes pure white in color; termed the Lord of the Deep by Polynesian fishermen. Sightings support the possible survival into modern times of the megalodon shark *Carcharocles megalodon*, which is known to have attained a total length of at least 43 feet (i.e. twice that of the great white), and may have survived until as recently as the Pleistocene's close (Ellis & McCosker, 1991; Shuker, 1995f, 1997d).

Strikingly marked manta ray, with a 10-foot wingspan. Predominantly dark brown and faintly mottled, but with conspicuous white wing-tips and a pair of broad pure-white bands extending halfway down the back from each side of its head. Spied by William Beebe and crew members of his vessel *Noma* on April 27, 1923, as they sailed toward Tower Island in the Galapagos Archipelago, when the ray collided with *Noma* and then sped off along the sea's surface. No known species of manta possesses its white markings, and its cephalic horns were more straight than incurved, unlike those of other mantas (Beebe, 1924). A somewhat similar manta was filmed in the 1980s off southern Baja California (Sehm, 1996; Shuker, 1999).

Planctosphaera pelagica, a mysterious acorn worm (enteropneust) species, whose adult form is still unrecognized by science. Its tornaria larva, resembling a transparent gooseberry-sized sphere and much larger than the tornaria of any known acorn worm, was first seen in 1932, when the *Michael Sars* Deep-Sea Expedition collected two specimens at a depth of 830 feet from the plankton of the North Atlantic's Bay of Biscay (Spengel, 1932; Shuker, 1993b).

Lophenteropneusts—name given to a taxon (conceivably a class) of hitherto-unknown and still uncollected invertebrates whose morphology resembles a combination of features from the two presently recognized classes of the invertebrate phylum Hermichordata, i.e. the acorn worms (Enteropneusta) and the pterobranchs (Pterobranchia). Lophenteropneusts were first photographed in 1962, during the PROA Expedition of the Scripps Institution of Oceanography, University of California, employing the research vessel "Spencer F. Baird," which took approximately four thousand photos of the sea bottom at hadal depths in five trenches in the South-West Pacific. Various of these photos depicted continuous coils and/or loops of fecal strings on the sea floor, some of which revealed an animal present at the end of the string. Measuring two to four inches long, these animals had a cylindrical hyaline body, with an anterior ring of tentacles, thus resembling pterobranchs, and a terminal anus, as in acorn worms (Lemche, et al., 1976).

Mysterious marine arthropod, never collected, but photographed several times at depths of approximately 13,616 feet in the Peru Basin on February 12, 1989, during the expedition DISCOL 1 (Disturbance and Recolonization Experiment in a Manganese Nodule Area of the Deep South Pacific). It possesses a small triangular fore-body measuring 0.52 inches long, and a hind-body measuring two inches long, separated from the fore-body by a narrow waist. The fore-body has five pairs of appendages, all jointed. The last three pairs seem to be long, slender walking legs. The first pair may function as feelers or assist in food uptake. The second pair is carried high above the body, mostly slightly bent downwards, and is even longer than the walking legs. This cryptid has been tentatively placed within the chelicerate class Arachnida, near to the amblypygids (Pedipalpi), and if correctly classified would constitute the first living species of deepwater marine arachnid (Thiel & Shriever, 1989).

Mystery squid represented only by paralarval specimens, described in 1991. Although only tiny, these paralarvae are very distinctive, with big eyes and possessing a pair of strikingly large lateral fins, on account of which this squid has been aptly dubbed "big-fin." Its taxonomic family, genus, and species are presently unknown. It may constitute the juvenile of an undescribed big-finned adult squid form videoed alive in several deepwater regions by submersibles during recent years but so far uncaptured (Young, 1991; Shuker, 2002b).

Unidentified, extra-large form of marine polychaete worm, some specimens of which may measure upward of 15 feet long, often observed by guests at the Anse Chastanet resort in Soufrière, St. Lucia, during night dives on the beach-access reef here, and popularly termed "The Thing." Partial specimens (all lacking the head) have been obtained. McGill University polychaete expert Dr. Susan Marsden deems it likely that "The Thing" belongs to the family Eunicidae. It could be a new, currently undescribed species, or an example of gigantism in a known species. A giant polychaete may also exist in the sea off Barbados (Roesch, 1996).

Colossal specimens of scyphozoan jellyfish, some with enormously long tentacles and typical bells, reported from Bermuda and the environs of Fiji; others of flattened shape, lacking tentacles, and of deepwater habitat, possibly including the legendary *hide* of Chile (Mangiacopra, 1976; Shuker, 1995f, 1997d).

FRESHWATER FORMS
(1) In Cold Temperate Lakes and Rivers
In Europe (Palearctic Region):

Aquatic mystery beast from Ireland known as the *dobhar-chú*, resembling a giant otter and also referred to as a master or king otter. Traditionally mythical, but one such beast is also claimed to have been the killer of a young woman called Grace Connolly, who was fatally attacked one morning in September 1722 while bathing in County Leitrim's Glenade Lake, near her home. Her grave still exists in Conwall Cemetery in the townland of Drummans, and on the tombstone is carved a distinctive otter-headed canine-bodied animal representing the *dobhar-chú*. Moreover, sightings of a mysterious quadrupedal beast closely resembling the creature sculpted on Connolly's tombstone were made near Sraheens Lough in Achill Island, County Mayo (to the west of County Leitrim), as recently as May 1968. Unless an unusually large version of the European otter *Lutra lutra* is involved, the *dobhar-chú* remains unexplained (Tohall, 1948; Shuker, 1995c, 1997d, 1998f, 1999).

In Cold Temperate Asia (Palearctic Region):

Giant salmon-like fishes, reputedly more than 30 feet long, and bright red in color with spiny dorsal fin, regularly spied in Lake Hanas, within China's Xinjiang Autonomous Region, by locals for several decades, and even by eminent Xinjiang University biologist Prof. Xiang Lihao during a student field trip there in 1985. Claimed by skeptics to be nothing more than the huchen (taimen) *Hucho taimen*, but this species does not exceed six and a half feet (Greenwell (ed.), 1986; Shuker, 1993b, 1997d).

In North Africa and Middle East (Palearctic Region):

Small but lethal black fish allegedly inhabiting Iran's Shatt al Arab River and currently unidentified by science, but said to have killed 28 people with a swift-acting venomous bite (Caras, 1975). If true, it could be related to *Meiacanthus nigrolineatus*, a small blue-grey and yellow species of freshwater blenny inhabiting the Red Sea and the Gulf of Suez and Aqaba. This fish does have a venomous bite, though it has not been shown to be fatal. Alternatively, it may be based upon confused reports of a small black catfish called *Heteropneustes fossilis*, which has poisonous spines in its pectoral fins, and which was fairly recently introduced into the Shatt al Arab River from Thailand and India as a food fish. There is even the possibility that young long-tailed moray eels *Thyrsoidea macrura*, swimming up the river from the sea, are implicated, as their bite can cause septic wounds (Shuker, 1996c).

In North America (Nearctic Region):

Very large type of salamander reportedly five to nine feet long, claimed to inhabit deep lakes in Trinity Alps, the Sacramento River, and elsewhere in California. Certain reports were based upon an escapee Chinese giant salamander *Megalobatrachus davidianus* called Benny. However, some authorities believe that others imply the existence of unrecorded populations of the American giant salamander (hellbender) *Cryptobranchus alleganiensis,* officially known only from the eastern U.S., or even an undiscovered American species of *Megalobatrachus* (Rodgers, 1962; Coleman, 1989; Shuker, 1995f). There are also reports of extra-large pink-colored salamander-like creatures, sighted in New Jersey and Florida, which may represent an albinistic morph of a giant salamander species (Hall, 1992; Shuker, 1995f).

(2) IN TROPICAL LAKES, RIVERS, AND SWAMPS
In Asia (Oriental Region):

Extra-large form of freshwater turtle, up to six and a half feet long, recently sighted and videoed in Hoan Kiem Lake, situated in modern-day central Hanoi, Vietnam, and also represented by a stuffed but unclassified specimen presently exhibited at a small temple on an island in this lake. According to Hanoi National University biologist Prof. Ha Dinh Duc, who is studying this sizeable freshwater chelonian, it is likely to prove to be a new species (Shuker, 1998g).

A six-foot-long species of lungfish in Vietnam, judging from excellent first-hand anecdotal reports collected by Dr. Roy P. Mackal (Shuker, 1991b).

Unidentified mudskipper-like fish observed in a river on the Indonesian island of Seram (Ceram) by tropical agriculturalist Tyson

Hughes in 1986, and which glows a bright pulsating red color at night. Hughes attempted to catch a specimen, but was unsuccessful (Shuker, 1995b, 1999; Gibbons, 1997).

In Africa (Ethiopian Region):

Greatly feared aquatic cryptid from Burundi known as the *mamba mutu*. Local eyewitnesses (mostly from villages on the shores of Lake Tanganyika and along the Lukuga River) describe it as half-human, half-fish, in the classic mermaid mold. However, they also firmly believe that it kills people, eats their brains, and sucks their blood. It has been suggested that a lake-dwelling population of African manatees, transformed by native superstition into vampiric monsters, may be the solution; or that an unknown species of giant flat-skulled otter is involved (Bonet, 1995).

In Central and South America (Neotropical Region):

Aquatic feline beasts with large canine teeth and yellowish, unpatterned pelage, reported from Patagonia *(yaquaru)*, Guyana *(maipolina)*, also called water-tigers. Often confused with otter-like *iemisch* and ground sloth-like *ellengassen*. Possibly a surviving machairodontid (sabre-toothed cat), secondarily adapted for an amphibious existence and thus paralleling Africa's so-called water-lions, water-panthers, and water-leopards (Shuker, 1989, 1995f).

Anomalous pink river dolphin, resembling the bouto (inia) *Inia geoffrensis*, except for its dorsal fin, which, instead of being triangular, possesses a regular series of notched projections, so that it resembles the edge of a circular saw. Seen, photographed, and filmed on two separate occasions during the mid-1990s by English writer/explorer Jeremy Wade in and near a lake beside one of the Amazon's countless southern tributaries in Brazil. Known to the native people and possibly constituting a rare freak of nature, as the precisely formed shape of the notches would seem to eliminate an injury-induced explanation for its "sawtooth" dorsal fin (Wade, 1996; Shuker, 1997f).

Huge toothless shark claimed by Lieut.-Col. Percy Fawcett to exist in the Paraguay River, and given to attacking and swallowing people if presented with the opportunity (Fawcett, 1953). Yet very few sharks inhabit fresh water, and those that do are far from toothless. Assuming, therefore, that it is not some dramatically new species of shark, it has been tentatively suggested that this aquatic cryptid may be a giant sturgeon (Grayson, 1998). More likely, however, is that it is a species of giant catfish, some of which are indeed toothless, and there are already more species of catfish documented from South America than from anywhere else in the world (Shuker, 2000).

TERRESTRIAL FORMS
(1) In Europe (Palearctic Region):

Hungarian reedwolf—mysterious canid of controversial identity often reported from Hungary and eastern Austria until early 1900s, but may now be extinct. Many believe that it is a small race of wolf, but others still favor an extra-large form of common jackal. Time for DNA analyses, utilizing museum skins? (Éhik, 1937-1938; Tratz, 1958; Shuker, 1991b).

Unidentified, brown-colored, burrow-dwelling lizard, one foot long, with orange-sized head, pronounced dewlap, and long tongue; known locally as the *cenaprugwirion*, or *genaprugwirion* ("daft flycatcher"), and reported from Abersoch, North Wales (Wallis, 1987). Presumably comprising a naturalized population of some non-native species of lizard, originating from accidental escapes of specimens from private collections into the wild. An extremely remote but tantalizing alternative is a naturalized population of New Zealand's famous lizard-like "living fossil," the tuatara *Sphenodon punctatus*—a very hardy, long-lived, burrow-dwelling species of comparable size and appearance, which was often maintained in captivity in Britain during the 19th century (Shuker, 1997b, 1997d).

Controversial population of chameleons discovered around Pylos Lagoon, Greece, by biologist-photographer Andrea Bonetti; studied by her since 1996. Larger than the European chameleon *Chamaeleo chamaeleon*, they bear a much closer (but not exact) similarity to the African chameleon *C. africanus*. If not conspecific with the latter, it is conceivable that these previously overlooked lizards represent a new species (Bonetti, 1998).

(2) In Cold Temperate Asia (Palearctic Region):
Extremely large, snow-white bear with small head and strange ambulatory gait, reported from Russia's Kamchatka peninsula and known to local reindeer hunters as the *irkuiem* or *irquiem*. Currently represented by at least one skin (exhibited at the St. Petersburg Museum), in the opinion of renowned Russian zoologist Prof. Nikolaj Vereshchagin it may be a modern-day representative of the Pleistocene short-faced (bulldog) bear *Arctodus simus* (Greenwell (ed.), 1987; Shuker, 1995f, 1997d).

Mysterious white bird referred to as Steller's sea raven, sighted by German naturalist Georg Steller on far-eastern Russia's Bering Island during the 1740s but apparently never seen again. Danish zoologist Lars Thomas considers it possible that this bird was a young (or even albinistic) spectacled cormorant *Phalacrocorax perspicillatus*—a species also discovered in this region by Steller but now extinct. Alternatively, it may have been some species now known to science but unfamiliar to Steller, such as the surfbird *Aphriza virgata*, nominated as one option by Steller scholar Chris Orrick (Shuker, 1999, 2000).

Vermiform mystery beast reportedly inhabiting the southern Gobi desert, known to Westerners as the Mongolian death worm, and to nomadic Mongolians as the *allghoi khorkhoi* ("intestine worm"), on account of its supposed resemblance to a cow's intestine. It looks like a fat dark-red worm, three to five feet long, with no differentiated head, tail, or limbs, and spends much of its time hidden beneath sands, but is occasionally spied resting on the surface during June and July. Avoided by humans, camels, and other large species, the death worm is so-named because it can reputedly squirt a lethal corrosive poison from one end of its body, and can also supposedly kill in a much more mysterious manner, resembling electrocution. It has been sought in recent years by Czech explorer Ivan Mackerle (Mackerle, 1996; Shuker, 1998a). If its lethal capabilities are merely the product of folklore, the death worm could be a giant species of amphisbaenid, or even (albeit much more radically) a highly specialized species of earthworm with a water-retentive cuticle. If its death-dealing abilities are real, however, it is more likely to be a dramatic new species of snake, possibly an elapid akin to the Australian death adder *Acanthophis antarcticus* (Shuker, 1998a).

(3) In North Africa and the Middle East (Palearctic Region):

Qattara cheetah of Egypt—reports date back to the late 1960s, and include a specimen taken by a Bedouin shepherd in 1967, but none observed by scientists so far, although cheetah-like tracks have been photographed in the Qattara region on two separate occasions during the 1990s. It is said to be paler in color and with a thicker coat than the typical cheetah from eastern and southern Africa. Probably similar to the little-known desert-adapted cheetahs with speckled, sandy pelage living in Niger's Tenere Desert, West Africa, and photographed for the first time as recently as the early 1990s (Ammann, 1993; Hoath, 1996).

(4) In Tropical Asia (Oriental Region):

While visiting the tiny Indonesian island of Rintja, close to Komodo, French traveler Pierre Pfeiffer collected native reports of a mysterious beast called the veo (Pfeiffer, 1963). At least 10 feet long, the veo has a long head, huge claws, and is covered dorsally and laterally with large overlapping scales, but has hair upon its head, throat, belly, lower legs, and the end of its tail. If threatened, it will rear up on its hind legs and slash its antagonist with the formidable claws on its forepaws. Inhabiting mountains during the day, but descending to the mangrove coasts at night, it lives upon ants, termites, and small sea creatures stranded on the beach. This description recalls the pangolins (scaly anteaters), but none of those species known to exist today is as big as the veo. During the Pleistocene, however, a giant species, *Manis palaeojavanicus*, over eight feet long, lived in Borneo and Java. Perhaps this species or a modern descendant persists today on Rintja (Mareš, 1997; Shuker, 1998h; Shuker, 1999).

While working for the VSO (Voluntary Service Overseas) in Seram, Indonesia, during 1986, tropical agriculturalist Tyson Hughes heard native accounts of a mysterious entity called the *orang bati* ("flying man"). Villagers in Seram's coastal regions live in terror of these creatures, which, they claim, live in long-dead volcanic mountains in the island's interior. At night, they leave their mountain lairs and fly across the jungles to the coast, where they seize babies and infants from the villages and carry them back to the mountains. According to eyewitness testimony, the *orang bati* is humanoid in form, with red skin, black wings, and a long thin tail. Perhaps these claims are based upon the existence here of a scientifically undiscovered species of giant bat, akin to the Javanese *ahool* (Shuker, 1995b).

Form of loris closely resembling slow loris *Nycticebus coucang*, but instantly distinguished from all known lorises by possessing a long bushy tail; indigenous to India's Lushai Hills. Specimens were captured alive, photographed, and exhibited for a time in captivity during 1889 (Annandale, 1908). On the assumption that this tailed loris represents a genuine species (rather than a teratological aberration), and based upon a concise published description and photograph (Annandale, 1908), it has been named *Nycticebus caudatus* sp. nov.

(Shuker, 1993b), but has not been reported in recent years, so may now be extinct.

Mysterious unidentified loris superficially resembling the slow loris but much larger and far paler in color, exhibited and photographed in a bird cage on an animal dealer's stall at the new animal market in Hanoi, Vietnam, in December 1994 (Richardson, 1995; Shuker, 1997d).

Nepalese cryptid with striped tiger-like body and canine head, known as the *chuti*, and often depicted in traditional art. Nepalese lamas informed mountaineer Hamish MacInnes that *chutis* inhabited the Choyang and Iswa Valleys. Russian scientist Dr. Vladimir Tschernesky has suggested that this cryptid may be the striped hyaena *Hyaena hyaena* (MacInnes, 1979).

Apparently unknown civet inhabiting Seram, Indonesia. While working here in 1986, Tyson Hughes collected the remains of a mammalian tail, about 18 inches long, which the native Moluccans claimed was from a large cat-like beast. It was later identified by WWF experts as the tail of a civet, but there is no known species of civet in Seram (Shuker, 1995b).

Sumatran mystery cat with unpatterned yellow or tan fur, a short tail, and a ruff encircling its neck, known locally as the *cigau*. According to native reports collected by British explorer Debbie Martyr, the *cigau* is slightly smaller but more heavily built than the Sumatran tiger, is greatly feared by the local people on account of its aggressiveness, and allegedly frequents the wilderness region east of Mount Kerinci and south toward the market town of Bangko (Shuker, 1995e; Heuvelmans, 1996).

Equally mystifying is the much smaller feline cryptid reputedly inhabiting the islands of Alor and Solor in the Lesser Sundas, southeast of Sumatra and Java. Judging from reports collected by Debbie Martyr, this cat is only the size of a domestic cat, but is readily distinguished from all other felids by the pronounced knob-like protuberances, resembling short stubby horns, present upon its eyebrows (Shuker, 1995e, 1999).

Sheepdog-sized, tiger-like felid reported from the Ryukyu island of Iriomote, where it is called the *yamamaya*. It could be an undescribed subspecies of tiger, or even clouded leopard (Anon., 1968; Shuker, 1989).

Another tiger-like mystery felid from Asia is the *seah malang poo*, supposedly inhabiting Thailand's Khao Sok National Park. Living in this area's karst limestone mountains, it is said to be of stocky build with brown and black stripes. A skin from one such cat, shot in the 1930s, was allegedly sent to Thailand's national museum, but further details regarding this specimen have not been forthcoming to date (Belderson, 1995; Day, 1995; Shuker, 1995e).

Small wolf-like canids persistently reported in Nara Prefecture and elsewhere in Japan, and photographed on October 14, 1996, at close range in the mountain district of Chichibu. They greatly resemble the shamanu or Japanese dwarf wolf *Canis lupus hodophilax*, whose last confirmed specimen was killed in Nara Prefecture during January 1905. Reports have been sufficiently convincing for the staging on March 19-20, 1994, of a two-day symposium devoted to the shamanu and its putative post-1905 survival. It is certainly likely that some reports are based upon misidentified feral dogs, and possibly even the descendants of original pre-1905 matings between shamanus and dogs, but certain reports on file do readily recall the shamanu's distinctive, diminutive morphology (Anon., 1997a; Shuker, 1997c, 1998d).

"Horned" jackals in Sri Lanka—occasional male specimens of the common jackal *Canis aureus* possessing a very small horny projection, usually at the rear of the skull. Horn development is no doubt controlled genetically, or induced epigenetically by physical injury (Tennent, 1861; Shuker, 1997d).

Supposed presence of tapirs in Borneo, including capture of an apparent juvenile specimen (dubbed a "tigelboat" by popular press) in Kalimantan (Indonesian Borneo) during 1975. Although probably conspecific with the Malayan tapir *Tapirus indicus*, which inhabited this island during the Pleistocene, they may comprise a distinct subspecies, awaiting formal description (Shuker, 1995f).

Porcine beast known by the Saiap Dusuns of North Borneo (Sabah) as the *pukau*, said to occur in large numbers on Mount Madalong, and to resemble a hybrid of pig and deer. This description is reminiscent of Sulawesi's babirusa *Babyrousa babyrussa*—which could conceivably have entered Borneo in pre-Holocene times across transitory land bridges existing during that period between Borneo and the faunal region of Wallacea, which includes Sulawesi (Rutter, 1929; Shuker, 1995f).

In 1994, Vietnamese biologist Nguyen Ngoc Chinh visited Pu Mat, an area just north of Vu Quang (scene of the discovery in 1992 of the Vu Quang ox *Pseudoryx nghetinhensis* and the giant muntjac *Megamuntiacus vuquangensis* in 1994) in northern Vietnam. While there, local hunters gave him the skull of a mysterious deer known to them as the *quang khem* or slow-running deer, but seemingly undescribed by science. Unlike those of other deer, the antlers of this strange animal are not branched, but simply comprise a pair of short pointed spikes, so that its skull resembles the horned helmet of a Viking warrior. Some additional *quang khem* skulls, previously neglected, have been uncovered by *Pseudoryx* and *Megamuntiacus* discoverer Dr. John MacKinnon, a British conservationist, in a box of unsorted bones at Hanoi's Institute of Ecology and Biological

Resources. DNA samples taken from these were analyzed by Copenhagen University geneticist Dr. Peter Arctander, who was unable to match them with the DNA of any known species. To date, however, no complete specimen of the *quang khem* has been spied by scientists or made available for scientific study (Linden, 1994; Shuker, 1995a, 1997d, 2002b).

Also discovered by MacKinnon in that same box of unsorted bones was a pair of antlers belonging to another species of deer seemingly unknown to science but familiar to the local Vietnamese hunters. They call this one the black deer or *mangden* (Shuker, 1995a, 1997d).

Possible survival of Schomburgk's deer *Cervus schomburgki* in Laos. Previously restricted largely to central Thailand's Chao Phraya Basin, and believed to be entirely extinct since the 1930s, this handsome species was famous for the male's extremely ornate, extensively branched antlers. In February 1991, a pair of antlers belonging to this species was spotted in a medicine shop in Laos by Laurent Chazée, a United Nations agronomist. When questioned regarding their provenance, the shop owner claimed that they were from a deer shot in a nearby district in 1990 (Schroering, 1995). With several remarkable new ungulates having already been discovered in Laos and neighboring Vietnam during the 1990s, it would be rash to deny the possibility that Schomburgk's deer also exists here, unseen by scientists.

Referred to as the holy goat or *linh duong* in Vietnam and as the *kting voar* in Cambodia, *Pseudonovibos spiralis* was formally described and named in 1994, but this enigmatic ungulate is still known to science only from its distinctive spiraled horns. Local hunters claim that it is grey-brown and resembles a cow, but no scientist has yet seen even a skin or skeleton of one, and some have denounced it as a hoax (Shuker, 1995d, 1997d, 2002b).

Possible persistence of the Indian pink-headed duck *Rhodonessa caryophyllacea*, believed extinct since the 1940s. Most eyewitness claims have emerged from regions currently difficult to explore due to political unrest (Shuker, 1991b). The possibility that this species survives in Tibet, which is outside its known distribution range, was investigated (uneventfully) by a British expedition during 1998 (Anon., 1998).

Large, possibly flightless species of argus pheasant, christened *Argus* [now *Argusianus*] *bipunctatus* in 1871, on the basis of a single feather that bore two bands of dots instead of just one (as in the great argus *A. argus*, the only *Argusianus* species currently recognized). Its provenance is unknown, but suggestions include Java and the Malaysian island of Tioman; it is probably now extinct (Davison, 1983; Shuker, 1990a, 1991a, 1999).

Chicken-shaped, chicken-sized gallinaceous bird known as the *alovot*, reported from Sumatra. It is said to have a comb-like crest in some instances (perhaps present only in one sex?), and dark brown plumage dappled with lighter spots. Too small to be the great argus, it could conceivably be an undescribed species of *Polyplectron* peacock-pheasant (Jacobson, 1937; Shuker, 1990a, 1991a, 1993b; Heuvelmans, 1996).

Long-tailed, pigeon-sized bird known as the *ulama* (Singhalese), *andai* (Tamil), and devil-bird (English), inhabiting forests of Sri Lanka. Rarely spied but often heard, its name derives from its hideous cry, described by one earwitness as resembling the sound that would be made by a boy being slowly strangled. Several identities have been postulated, including one or more species of the island's known owls, raptors, rails, and nightjars (Tennent, 1861; Shuker, 1991b, 1997d).

An unidentified owl species belonging to the genus *Strix* has been reported on more than one occasion from the Andaman and Nicobar islands (Ali & Ripley, 1969).

Minute (one and a half inches long) bird resembling the New World hummingbirds but reported in Sumatra. Two specimens were seen flying together in c.1957-1958 at a distance of only 12 inches away from the face of one local eyewitness and his wife. They were yellowish in color, with dark brown bellies and little stripes (Coleman, 1989). The hummingbirds are exclusively New World in distribution, but there are Old World passerines that are superficially similar, known as sunbirds (nectariniids). None currently known to science, however, is as small as the birds described here. Although unlikely, perhaps the eyewitnesses mistook a species of nectar-seeking day-flying moth for a bird—as often happens in Britain, for instance, when the tiny summer-visiting hummingbird hawk moth arrives from the European continent.

(5) In Tropical and Southern Africa (Ethiopian Region):
Small, unidentified microchiropteran bat from Kenya, distinguished by its unique daytime hiding place—piles of dry elephant dung. It may prove to be the little-known species *Rhinopterus floweri*, or an undiscovered relative (Williams, 1967; Shuker, 1994).

Sanguinivorous (blood-drinking) microchiropteran bat allegedly inhabiting Devil's Cave in Walaga Province, southern Ethiopia, and termed the death bird by local peasants. As the only sanguinivorous chiropterans are exclusively Neotropical (the vampire bats), if reports are genuine it is likely to constitute either a totally new species, or a known species that has unexpectedly acquired a sanguinivorous lifestyle in this particular geographical location (Prorok, 1943; Shuker, 1994, 1997d).

Cat-sized bushbaby-like primate observed and photographed in Cameroon by an assistant of bushbaby taxonomist Dr. Simon K.

Bearder in 1994. It is most likely an extremely large new species of bushbaby, but also worth considering is the exciting possibility that it constitutes the first living species of true lemur recorded from mainland Africa (Bearder, 1997).

Unidentified tiny lemur reported from the Tsingy de Bemaraha nature reserve in western Madagascar. Known to the Sakalava du Menabe tribe living here as the *malagnira,* and claimed by them to be similar to but even smaller than the *Microcebus* mouse lemurs (the world's smallest known lemurs), and with different behavior (Rakotoarison, et al., 1993). Some notably bigger new species of lemur have been discovered and rediscovered in Madagascar during the past two decades (Shuker, 2002b), so the putative reality of a much smaller species still eluding scientific detection cannot be discounted.

Possible 20th century survival of the officially long-extinct giant aye-aye *Daubentonia robusta,* as indicated by the discovery in c.1930 of an exceptionally large aye-aye skin at a native's home near the village of Andranomavo in Madagascar's Soalala District by a government official called Hourcq (Hill, 1953).

Oliver—a controversial male chimpanzee, attracting public attention in 1970s due to his distinctive appearance and ability (seemingly unforced) to walk bipedally erect; now living at Primarily Primates, an animal sanctuary in Texas. Variously identified as an unexceptional but well-trained West African chimp, a mutant chimp, a hybrid of common and pygmy chimp, a new chimp (sub)species, a new species of non-chimp primate, and even a hybrid of chimp and human, or of pygmy chimp and a West African bipedal crypto-primate called the *séhité.* The last two identities were proposed in response to an unsubstantiated claim that Oliver has 47 chromosomes (one more than humans, one less than chimpanzees); in reality, he has 48. Moreover, chromosomal studies and mtDNA sequence analyses conducted by a team of American geneticists including Drs. Charleen Moore and John Ely have revealed that Oliver has a normal chimp karyotype and that he closely resembles a Gabon chimp of the subspecies *Pan troglodytes troglodytes* genetically (Anon., 1976, 1997b, 1997c; Ely, et al., 1998; Ely & Moore, 1998).

Highly aggressive, southern Cameroon primate known to Baka pygmies and Bantus as the *dodu*. According to explorer Bill Gibbons, they claim that it is dark grey, stands up to six feet tall, is mostly bipedal but sometimes knuckle-walks on all fours, and, very distinctively, has only three fingers on each hand and just three clawed toes on each foot. It will attack gorillas, and leaves piles of sticks on the forest floor, possibly a form of territorial marking behavior (Shuker, 2003).

Large African leopard-like felids but with black upperparts and paler underparts, reported widely through tropical Africa but particularly in Uganda *(ndalawo)* and the Aberdares of Kenya *(damasia)*. Judging from their morphology, they seem likely to be pseudomelanistic leopards (like the "melanotic" specimens recorded from Grahamstown, South Africa, during the late 1800s), as opposed to melanistic leopards (i.e. black panthers) (Shuker, 1989).

Large African felids with dark blotches or stripes, reported from Ethiopia *(wobo)* and Sudan *(abu sotan)*. The *wobo* skin formerly held at the cathedral of Eifag may have been an imported tiger skin (Shuker, 1989).

Mysterious giant carnivorous bird, capable of flight yet allegedly even taller than the ostrich, claimed during the 1870s by the Wasequas to inhabit their country, situated on the African mainland eight to nine days' journey away from the coast of Zanzibar. Said to feed primarily upon carrion, it is called the *makalala* ("noisy"), on account of the loud noise produced when it claps its wings togeth-er, which possess horny plates at their tips. It also has long legs, and the head and beak of a bird of prey. Allowing for exaggeration in relation to its height, this description irresistibly calls to mind an extra-large version of the secretary bird *Sagittarius serpentarius*, which even possesses horny tips to its wings. Accordingly, and based upon a published description of the *makalala* (Marschall, 1878-1879), it has been named *Megasagittarius clamosus* gen. nov., sp. nov. ("noisy giant secretary bird") (Shuker, 1995f). No modern-day records of the *makalala* are known, so even if it did exist and was a giant species of secretary bird, it is assuredly extinct today. However, its skulls were said to be highly prized as ceremonial helmets, worn only by chiefs. Perhaps, therefore, it is not too late to seek out one of these skulls, and thereby solve the mystery of this avian cryptid's identity (Marschall, 1878-1879; Shuker, 1995f, 1996d).

Glossy-black fowl-like bird with red bill and claws, no differences between sexes, and voiceless; flies low, like a guineafowl, is called the *kondlo*, and is native to Zululand, South Africa. Although it is believed by some to be either the southern ground hornbill *Bucorvus leadbeateri* or the bald ibis *Geronticus calvus*, experienced Western hunters familiar with those latter birds and with the *kondlo* disagree with these claims. It is possibly an undescribed galliform bird (Court, 1962; Shuker, 1991a, 1996d, 1999).

Unidentified form of stone partridge *Ptilopachus petrosus* native to Senegal. Distinguished ecologically from the known, nominate sub-species here by inhabiting forests and dense undergrowth rather than this species' typical rocky terrain or scrub. Distinguished mor-phologically by its spotted head, paler breast, and smaller size (Shuker, 1991a).

Undescribed species of gallinule inhabiting the Sudd. Said to be noc-turnal (Kingdon, 1990).

Unidentified green touraco with a very little red upon its wings, never collected but spied by at least three eminent ornithologists, including John G. Williams, within the Impenetrable Forest in south-western Kigezi, Uganda (Williams & Arlott, 1980). African wildlife expert Dr. Jonathan Kingdon deems it likely that this bird is based upon fleeting glimpses of the scarce southern race of Ruwenzori touraco *Tauraco johnstoni*, which sometimes flies in a fast short flight, exposing little of its red wings (Kingdon, 1990).

Very large, unidentified all-black swift, observed on Marsabit Mountain in Kenya's North Frontier Province (Williams & Arlott, 1980).

Greyish, long-tailed mystery bird with red or chestnut under tail-coverts, spied in Kenya's Matthews Range (Williams & Arlott, 1980). Kingdon (1990) suggests that this may be the chestnut-winged star-ling *Onychognathus morio*, noting that if the bird is not flying it is easy to confuse a glimpse of the primaries below the tail with the under tail-coverts.

Bizarre-sounding reptile called the *das-adder*, reported from South Africa's Drakensberg Mountains, and said to have a viperine body but a hyrax-like head. Stripping away the layers of folklore undoubtedly encapsulating this beast, it is likely that its true identity will prove to be one of the region's larger monitor lizards (Speight, 1940; Shuker, 1997d).

Equally bizarre mystery snake reported from southern Namibia's Namib Desert by the Namaqua people and also by various Western eyewitnesses. It reputedly possesses a huge head, inflated neck, a yellow or brown body speckled with dark spots, and a pair of bat-like wings emerging from the sides of its mouth or neck, plus a pair of short backward-curving horns on its head and—most amazing of all—a bright glowing light in the centre of its brow (Whitty, 1995;

Muirhead, 1995). According to eyewitnesses, it can launch itself from the summit of a high rocky ledge, and soar down to the ground, landing with a considerable impact, and leaving behind scaly tracks in the dusty earth (one such track was examined by coelacanth discoverer Marjorie Courtenay-Latimer, who confirmed that it resembled a snake's track). Assuming that this is an exaggerated account of a genuine snake (perhaps influenced by Germanic lightning snake legends and folklore reaching Namibia via the arrival here of German settlers), it may in reality be an undescribed species possessing a pair of extendable lateral membranes comparable with those of the famous Asian gliding lizard *Draco volans*, and a highly reflective patch of shining scales on its brow (Shuker, 1996b, 1999).

Undiscovered species of hawk moth possessing a proboscis at least 15 inches long, and inhabiting Madagascar. The existence of such an insect was predicted in 1992 by Ohio entomologist Dr. Gene Kritsky as the only explanation for the successful pollination of a certain species of Madagascan orchid, *Angraecum longicalcar,* whose nectar-producing organs (nectaries) are 15 inches deep inside the plant, and hence can only be reached by a moth with a suitably lengthy proboscis (Angier, 1992). A notable precedent for this case was the formal description in 1903 of *Xanthopan morgani praedicta,* a Madagascan hawk moth with a 12-inch-long proboscis. Its existence had been correctly predicted back in 1862 by Charles Darwin, who recognized that such a moth must certainly exist in order to explain the successful pollination of the orchid *Angraecum sesquipedale*—whose nectaries are roughly 12 inches deep, and hence beyond the reach of any Madagascan moth known to science at that time (Shuker, 1993b).

(6) In North America (Nearctic Region):
Regular sightings of puma-like cats in the eastern United States, where the eastern cougar *Felis concolor couguar* officially became

extinct several decades ago. Some such sightings, even if genuinely featuring pumas, could involve specimens of other subspecies
that have escaped or have been released from captivity. Some may
even feature western subspecies that have migrated into the east,
occupying the ecological niche left vacant by the disappearance of
F. c. couguar. Bearing in mind, however, that the puma is highly
skilled in concealment, it would not be impossible for a small con-
tingent of bona fide eastern cougars to have persisted within their
native distribution range (Shuker, 1989; Tischendorf & Ropski, 1996).

Occasional reports of extremely large spotted mystery cats in
southern U.S., probably involving absconded leopards and jaguars
from captivity, but may also indicate survival of the Arizona
jaguar *Panthera onca arizonensis,* officially extinct since 1905 (Shuker,
1989).

Very large, all-white wolf-like beast with unusually wide paws and
head, small ears, rather short legs, and solitary habits, reported from
northern Alaska and Canada's Northwest Territories, and known to
the native Indians as the *waheela.* It has been suggested that it could
comprise a surviving member of the amphicyonids or bear-dogs—
a family of burly, plantigrade, bear-like carnivores with lupine dentition and face, currently assumed to have died out in North America
around two million years ago, and elsewhere by the end of the
Pleistocene epoch, 10,000 years ago (Sanderson, 1974; Shuker, 1995f).

Anomalous cat-sized mammal said to resemble a goat, with rose-
colored horns and snow-white silky fur, but clawed feet. One was
reputedly owned by an Indian woman, and was seen at
Fredericksburg, Texas, around the mid-1800s by an American officer who described it to the Abbé Emanuel Domenech (Domenech,
1858; Shuker, 1997d).

Not all reports of living thunderbirds or "big birds" in North

America fit the feathered, vulturine type popularly assigned a ter-
atorn identity by cryptozoologists. Some, reported from Texas
and Pennsylvania, seem to be featherless and much closer in
appearance to giant fossil pterosaurs, notably the North American
Pteranodon and *Quetzalcoatlus*. The prospect of pterosaurs persisting
beyond the Cretaceous into the present day in this continent must
surely seem highly unlikely. Even so, it is a formidable coincidence
that regions in which the fossils of such creatures have been
unearthed had in some cases already hosted sightings of cryptids
bearing a close resemblance to the likely appearance in life of these
same fossil species (even allowing for the transforming effects of
more than 65 million years of continuing evolution) (Clark &
Coleman, 1978; Shuker, 1995f).

(7) In Central and South America (Neotropical Region):
Likelihood that at least one species of the West Indies' large, supposedly long-extinct family of nesophontid insectivores still survived as lately as the 1920s (and perhaps persisting even today);
based upon discovery of seemingly fresh skeletal remains extract-
ed from recent owl pellets on Hispaniola (Miller, 1930; MacFadden,
1980; Shuker, 1993b).

Unusually large vampire bats allegedly attacking cattle and horses, reported by locals from southeastern Brazil's Ribeira Valley—
which is also a source of Pleistocene remains from an unusually
large, officially extinct species of vampire bat, *Desmodus draculae*.
Coincidence, or evidence for Recent survival of *D. draculae*? (Trajano
& Vivo, 1991; Shuker, 1995f).

Chimpanzee-sized arboreal monkey with a baboon-like face and a
very short tail, called the *isnachi* and other local names by Indians
sharing its mountainous cloud-forest home in Peru. Said to be rare,
but extremely fierce, and is avoided by Indians whenever possible.
One of its most characteristic activities is to rip apart the tops of

chonta palm trees to obtain the tender vegetable matter inside—and as no other animal is strong enough to do this, its presence in a given locality can be confirmed by finding trees damaged in this way (Hocking, 1992).

Short-tailed Mexican mystery cat termed the ruffed cat, known from three reported skins, brown with wavy stripes on the flanks and upper limbs, one skin measuring just over seven and a half feet in total length, and all three skins characterized by a very conspicuous ruff of fur encircling the neck and covering the ears from above and behind. Two of these skins were supposedly purchased by Ivan T. Sanderson in 1940 from locals inhabiting a mountain settlement in the Mexican state of Nayarit's Sierra mountains, but both were subsequently destroyed when the government jail in Belize, where Sanderson had stored them, was flooded out. Sanderson claimed that he later saw a third such skin for sale at a market in Colima, at the south end of Nayarit's mountain block, but at a price above what he could afford to pay (Sanderson, 1973; Shuker, 1989).

Alleged new jaguar species spied but not caught by Dutch zoologist Dr. Marc van Roosmalen while spending over a year seeking new monkeys in Brazil during 1996 and 1997 (Anon., 1997d; Shuker, 2002b).

Jaguar-sized cat with tan fur and tigerine stripes, from Peru's Ucayali and Pasco provinces. It may be conspecific with striped mystery cats reported from Colombia and Ecuador. In 1993 Peruvian zoologist Dr. Peter Hocking obtained the skull of a female specimen, which will be studied by felid specialist Dr. Steven C. Conkling (Hocking, 1992, 1996; Shuker, 1996a).

Jaguar-sized cat, but with larger head, and grey fur covered by solid black speckles instead of the normal jaguar's golden fur patterned with black ring-like rosettes. Dubbed the "speckled tiger" by

Hocking, it is known to the Amuesha native hunters, and inhabits the lower Palcazu River valley in Peru's Pasco province (Hocking, 1992; Shuker, 1996a). In 1993, Hocking obtained the skull from a strange cat whose body size and fur patterning seem similar to the speckled tiger's, but its background color was said to be cinnamon-brown and white (instead of grey); its skull shows some differences from ordinary jaguar skulls. Hocking terms it an "anomalous jaguar" (Hocking, 1996). Another mystery cat reminiscent of the speckled tiger has been reported from Guyana and Brazil, and is called the *cunarid din* by the Wapishana Indians, who consider it to be an abnormal type of jaguar (Brock, 1963).

Mystery Peruvian cat resembling a giant black panther, entirely black and at least twice the size of the jaguar. Known to the Quechua Indians as the *yana puma* ("black puma"), it is said to be very aggressive at night, tracking hunters to their camps to kill them while they sleep (Hocking, 1992). Known melanistic pumas are extremely rare and are not uniformly black; their underparts are slaty grey. It could be an unusually large variety of melanistic jaguar, one whose cryptic coat markings are wholly concealed by its fur's black background color.

Unidentified lion-sized "jungle lion" with long hair around its neck and reddish-brown fur, reported by rangers in the Yanachaga National Park, Pasco, Peru (Hocking, 1996).

Pygmy brown bear sighted by rangers in Peru's Yanachaga National Park, and distinguished by them from the spectacled bear *Tremarctos ornatus* also existing here (Hocking, 1996).

Dog-like cat, or cat-like dog, called the *mitla*, reported from Bolivia by Lieut.-Col. Percy Fawcett and sought during 1960s by Jersey Zoo's director, Jeremy Mallinson. It may be the bushdog *Speothos venaticus*, or the small-eared dog *Atelocynus microtis*—a notably feline canid of uncertain distribution range (Fawcett, 1953; Mackal, 1980; Shuker, 1989, 1996a).

Alleged new tapir species seen but not captured by Dutch zoologist Dr. Marc van Roosmalen while spending over a year seeking new monkeys in Brazil during 1996 and 1997 (Anon., 1997d; Shuker, 2002b).

Apparently undescribed flamingo called the *jetete* in the Chilean Andes; distinguished by local inhabitants from this region's three known species (Walters, 1980; Shuker, 1993b).

Unidentified black, wattle-lacking guan in Peru's Yanachaga National Park (Hocking, 1996).

Small, mystifying sea-green macaws reported (even photographed) in various bird collections. Assumed to be freak Lear's or hyacinth macaws, but suspected by some to be specimens of the glaucous macaw *Anodorhynchus glaucus*, deemed extinct since 1930s (Shuker, 1993b).

Medium-sized snakes adorned with cockscombs and wattles, and gifted with the ability to crow like a cockerel—thereby paralleling similar mystery snakes in Africa—reported from Hispaniola and Jamaica in the West Indies (Gosse, 1862; Shuker, 1991b).

Extremely large form of batrachian, known as the *sapo de loma* ("toad of the hill"), reported from a number of Andean valleys in Chile and Peru. It is alleged to be very poisonous, and able to capture decent-sized birds (Shuker, 1997d).

(8) In Australasia (Australasian Region):

Out-of-place Tasmanian devils *Sarcophilus harrisii* reported and even occasionally captured on mainland Australia, including Victoria and

Western Australia, during the 20th century. These have most probably absconded from captivity (though none of the captured specimens has ever been traced back to such a source). Even so, as this
or a closely related species still occurred as recently as 600 years ago
in Victoria and a mere 400 years ago in southwestern Australia,
there is also a remote chance that *Sarcophilus* survives on the
Australian mainland even today, undetected by science (Troughton,
1965; Nowak, 1991; Shuker, 1998e).

Elusive dog-like cryptid with distinctly canine head, and stripes on
its back and tail, greatly resembling thylacine *Thylacinus cynocephalus*. Recorded regularly from Tasmania, where the last confirmed thylacine specimen died in 1936; from mainland Australia,
where the thylacine supposedly became extinct around 2,500 years
ago; and from mountainous regions of Irian Jaya (western,
Indonesian New Guinea), where natives refer to it as the *dobsegna*
(the thylacine is known from New Guinea via Pleistocene fossils).
The sheer quantity, and also the quality of so many, of the sightings
on record is such that after disregarding dingos, foxes, domestic
dogs, and other sources of misidentification, it is still reasonable to
suppose that thylacine survival beyond its respective accepted
extinction dates in all three regions has occurred. Its continuing survival in the more remote, inaccessible areas of Irian Jaya must be the
likeliest prospect, but its persistence in the other two regions
should not be ruled out (Shuker, 1993b; Healy & Cropper, 1994;
Shuker, 1995f; Smith, 1996).

Australian feline cryptid, as large as a clouded leopard, very
fierce, arboreal, with cat-like head, very prominent tusk-like teeth
at the front of its mouth, and distinctive black (or dark grey) and
white bands around its body. Mostly reported from forested regions
of northern Queensland, it is known to the aboriginals as the *yarri*,
to Westerners as the Queensland tiger, and can be readily delineated
from reports of dog-headed thylacine-like beasts. Judging from eye-

witness descriptions, it bears a striking resemblance to the predicted appearance in life (based upon reconstructions drawn from fossil evidence) of the marsupial lion *Thylacoleo carnifex,* which supposedly became extinct several millennia ago. Hence it has been suggested that the *yarri* is a surviving descendant (Heuvelmans, 1958; Shuker, 1989, 1995f).

Nocturnal Australian cryptid, superficially ape-like but allegedly distinct from Australia's famous man-beast, the *yowie.* Principally quadrupedal but can stand erect upon its hind legs and thus attain a height of five to six feet; it is equipped with very long muscular arms and elongated toes but lacks a tail. It is covered in dark fur upon much of its torso but lighter on its neck, limbs, and belly, and is known locally in New South Wales as the *yahoo.* Identities proposed include a surviving species of giant wombat, or a modern Australian representative of New Guinea's mountain diprotodont *Hulitherium thomasettii,* a marsupial with unexpectedly ape-like characteristics, believed extinct since the Pleistocene (Greenwell, 1994a; Shuker, 1995f).

Undescribed speckled megapode called the *sasa,* formerly native to the Fijian islands of Viti Levu and Kandavu, but probably now extinct (Wood, 1926; Shuker, 1991a).

Possible persistence of the huia *Heteralocha acutirostris* in the Kaimanawa mountain range of New Zealand's North Island. Presumed extinct since 1907, convincing sightings have continued to emerge of this very distinctive bird, famous for exhibiting marked sexual dimorphism in beak shape—short and pointed in the male, long and curved in the female (Phillipps, 1963; Shuker, 1991b; Greenwell, 1994b).

Six "lost" birds of paradise from New Guinea and outlying islands. Initially classed as valid species, they have been dismissed as hybrids since the 1920s, but recent research suggests that they may truly constitute genuine (albeit exceedingly rare) species, none of which has been reported for many years. Some or all of them may now therefore be extinct. As originally named, the six are: Elliot's sicklebill *Epimachus ellioti*, Sharpe's lobe-billed riflebird *Loborhamphus ptilorhis*, Rothschild's lobe-billed bird of paradise *L. nobilis*, Duivenbode's riflebird *Paryphephorus duivenbodei*, Bensbach's bird of paradise *Janthothorax bensbachi*, and Ruys's bird of paradise *Neoparadisea ruysi* (Fuller, 1979, 1995).

Black-plumaged, long-tailed mystery bird whose call sounds like a short explosive rattle, reported (but never collected) during the 20th century by various scientific teams visiting Goodenough Island— largest of the three main islands comprising the D'Entrecasteaux Archipelago just north of New Guinea's eastern tip. Suggested identities include a species of astrapia (long-tailed bird of paradise), a relative of the paradise crow *Lycocorax pyrrhopterus*, or a honeyeater (meliphagid) (Beehler, 1991; Shuker, 1999).

Unidentified winged beast with three-to-four-foot wingspan, long toothy beak, and lengthy tail terminating in a diamond-shaped flange, reported from island of Rambutyo (=Rambunzo) off the east coast of Papua New Guinea (PNG), and also from the isle of Umboi, between eastern PNG and New Britain. It is known locally as the *ropen*, and its morphology readily calls to mind that of *Rhamphorhynchus*, a small early prehistoric pterosaur (Shuker, 2002a).

A much larger winged mystery beast, often confused with the *ropen* by previous writers, reputedly inhabits mainland PNG, sporting huge leathery wings spanning up to 20 feet, a fairly long neck, and a bony crest. Known as the *duah*, it is seemingly reminiscent of *Pteranodon*, the mighty pterosaur of Mesozoic North America. Missionaries in the 1990s claim to have spied such creatures, and allege that they have glowing underparts (Shuker, 2002a).

Delcourt's giant gecko *Hoplodactylus delcourti* is currently known from a single stuffed specimen that had resided in Marseilles Natural History Museum for over a century before its species was recognized during the early 1980s to be unknown to science, and was formally described in 1986. The specimen's provenance is unknown, but was probably New Zealand, as its species' closest relatives are mostly endemic to New Zealand. Also, Maori mythology refers to a two-foot-long forest-dwelling lizard called the *kaweau* or *kawekaweau*, as thick as a man's wrist, and brown in color with red longitudinal stripes—a description very closely matching that of *H. delcourti*. In recent years, reports of lizards resembling this species have emerged from North Island, offering hope that this seemingly lost species may exist here (Hellaby, 1984; Bauer & Russell, 1987; Shuker, 1993b; Heuvelmans, 1996).

Unidentified aquatic mystery snake, reputedly very venomous, reported from PNG. No more than six feet long, with smooth scales, enlarged ventrals, and a short tail, one of these snakes bit three children near the village of Wipim, in Papua's Western Province, during 1972 and 1973; all three quickly died. It is said to favor small freshwater swamps and inland streams rather than rivers or open swampy grassland. English herpetologist Mark O'Shea has searched in vain for this snake during several visits to New Guinea, and cannot reconcile its description with that of any species known by science to exist here (O'Shea, 1996).

ADDENDUM: CHECKLIST AMENDMENTS AND ALTERNATIVE IDENTITIES FOR CERTAIN CRYPTIDS IN LIST

Heuvelmans's longstanding classification of sea serpents into nine well-defined types (Heuvelmans, 1968, 1986) has been challenged in a recent extensive critique (Magin, 1996). Its author concludes that none of Heuvelmans's postulated sea serpents will be discovered, because they do not exist. Instead, he proposes that they are merely cultural stereotypes: "the different types of creatures are based on patterns of speech, traditional metaphors of the region, cultural patterning and expectation" (Magin, 1996). Some of Heuvelmans's sea serpent types have also faced specific challenges from suggested alternative identities.

As discussed elsewhere (Shuker, 1995f), based upon recent paleontological discoveries and theories the present author considers that a plesiosaur identity offers greater morphological and behavioral correspondence with long-necked sea serpents (and their freshwater counterparts) than does Heuvelmans's hypothetical long-necked seal *Megalotaria longicollis*.

Heuvelmans nominated an armored, scaly archaeocete as the identity of his many-finned sea serpent *Cetioscolopendra aeliani* (Heuvelmans, 1968). However, it is now known that scales found in association with certain specimens of fossil archaeocete did not originate from them (as originally assumed), but belonged instead to various other creatures. In other words, there are no verified specimens of armored archaeocetes in the fossil record, thereby greatly reducing the likelihood of any modern-day species existing (Shuker, 1995f).

One of the most significant reports supporting the reality of Heuvelmans's Scandinavian super-otter *Hyperhydra egedei*

(Heuvelmans, 1968) originated with a sighting off the coast of Greenland made on July 6, 1734, by a priest, Hans Egede (in honor of whom Heuvelmans named the super-otter) and his son Poul. However, Danish zoologist Lars Thomas recently published, for the first time in any cryptozoological document, the original description of this creature, as penned separately by Hans and Poul Egede (Thomas, 1996), which reveals the creature to be very different morphologically from the super-otter envisaged by Heuvelmans. As a result, Thomas concludes: "It is so different, in fact, that I would suggest it is an entirely different animal to the super-otters Heuvelmans records as having been seen along the coast of Norway. Perhaps they should then be called the Norwegian super-otter *(Hyperhydra norvegica)*; certainly not Egede's super-otter" (Thomas, 1996).

A giant undiscovered species of swamp-dwelling lungfish inhabiting the Apa Tani's remote upland valley in northern Assam, India, has been nominated as a more compatible alternative to a monitor lizard for the identity of the *buru* (Izzard, 1951; Shuker, 1991b).

In his original checklist, Heuvelmans claimed that the *minhocão* was presumably an amphibious burrowing mammal (Heuvelmans, 1986), and had previously specified a surviving glyptodont as a contender for this identity (Heuvelmans, 1958). However, as proposed by the present author, an extremely large species of caecilian would provide a much closer morphological and behavioral correspondence with this cryptid (Shuker, 1995f); Heuvelmans also favored a fossorial amphibian in his updated checklist (Heuvelmans, 1996).

Heuvelmans's list opined that the *migo* of Lake Dakataua in New Britain may be an unknown crocodile or even a surviving mosasaur (Heuvelmans, 1986). During a Japanese expedition to this lake in 1994, with Dr. Roy P. Mackal as scientific advisor, a *migo* was filmed, leading Mackal to speculate that it may be a surviving, evolved archaeocete. Since participating in a second expedition here, however, during which time a clearer *migo* film was obtained, Mackal is convinced that the *migo* is not a cryptid at all—nor even a single animal. Instead, it is a composite, an optical illusion actually featuring three saltwater crocodiles *Crocodylus porosus* engaged in a mating ritual, involving two males and one female (Shuker, 1998b).

At least one type of British mystery cat is mysterious no longer— the Kellas cat of northern Scotland. This predominantly black, gracile, large-fanged felid, represented by several preserved specimens dating from the 1980s onwards, has lately been shown to be an introgressive hybrid of the Scottish wildcat and domestic cat— thus confirming the present author's predictions as to its likely taxonomic identity (Shuker, 1989, 1990b, 1993b).

In addition to the various identities offered by Heuvelmans as explanations for the Nandi bear in his checklist, the present author considers that two supposedly extinct mammals—the short-faced hyaena *Hyaena brevirostris* and a species of chalicothere—may also be involved (Shuker, 1995f). A chalicothere identity for this cryptid is supported by Brown University ungulate expert Prof. Christine Janis too (Janis 1987).

Heuvelmans's list offered a surviving deinothere as a putative identity for the Congolese water elephant (Heuvelmans, 1986). As noted by the present author, however, the latter cryptid shares little similarity with deinotheres, but it does greatly resemble the predicted appearance of a large yet otherwise little-changed, modern-day descendant of early proboscideans, such as *Phiomia* and *Moeritherium* (Shuker, 1995f).

Janis has suggested that the mysterious southern Ethiopian "deer" listed by Heuvelmans is much more likely to be a surviving

Ethiopian subspecies of the fallow deer *Dama dama* than a persisting representative of the antlered, outwardly deer-like giraffe *Climacoceras* as proposed by Heuvelmans, who erroneously referred to *Climacoceras* as a deer (Heuvelmans, 1986; Janis, 1987).

Researchers conducting biochemical and genetic tests upon tissue samples obtained from the carcass of the Rodriguez onza—the only onza specimen made available for full scientific examination—have so far been unable to differentiate it from pumas. They have therefore opined that this supposed onza is indeed merely a puma (Dratch, et al., 1996). This identity had already been predicted for it by the present author, who considers that onzas may well prove to be a gracile mutant form of the puma, rather than comprising either a distinct puma subspecies or an entirely separate species (Shuker, 1998c, 2002b).

The Brazilian *mapinguary* was listed by Heuvelmans as an unknown primate. However, Goeldi Museum zoologist Dr. David Oren has been conducting field researches in the dense Mato Grosso region of Amazonia concerning this cryptid for several years, and considers that it is much more likely to be a surviving mylodontid ground sloth. He has revealed that this region's Indians claim the *mapinguary* has red fur, is as tall as a man when squatting on its hind legs, leaves strange footprints that seem to be back-to-front and horse-like fecal droppings, emits loud shouting cries, and is said to be invulnerable to bullets. Although mylodontids officially died out a few millennia ago, they are known to science not only from fossils and preserved fecal droppings (which are indeed horse-like), but also from some remarkable mummified specimens, still covered with reddish-brown fur. These have revealed that mylodontids had bony nodules in their skin that would have acted as an effective body armor — thus explaining the *mapinguary's* invulnerability? (Oren, 1993; Shuker, 1995f).

In his original checklist, Heuvelmans commented that the outsized "rabbits" reported from central Australia may be giant kangaroos belonging to the fossil genus *Palorchestes* (Heuvelmans, 1986). However, as pointed out by Janis, it is now known that *Palorchestes* was not a giant kangaroo, but a genus of diprotodont. Consequently, Janis considers that a much more likely identity for Australia's giant "rabbits" is a surviving species of sthenurine kangaroo. Nevertheless, she does favor a modern-day species of *Palorchestes* (rather than *Diprotodon*, as suggested by Heuvelmans) as a plausible identity for the superficially tapir-like *gazeka* (more commonly termed the devil pig) of Papua New Guinea (Janis, 1987). These amendments were incorporated in Heuvelmans's updated checklist (Heuvelmans, 1996).

TABLES SUMMARIZING THE TAXONOMY OF CRYPTIDS LISTED IN THIS SUPPLEMENT

In his checklist, Heuvelmans included three tables, which classified and numbered the marine, freshwater, and terrestrial cryptids listed by him. The total number of cryptids in each category was as follows. Marine: 21-24; freshwater: 19-28; and terrestrial: 70-85. The reason why the totals were not precise is because in some cases Heuvelmans offered more than one identity for a given cryptid, and some cryptids listed by him may be composites, i.e. comprising more than one species erroneously lumped together as one by eyewitnesses.

I have provided three corresponding tables classifying and totaling the marine, freshwater, and terrestrial cryptids listed in this supplement. The totals given in these tables are precise. This is because I have only taken into account the most likely identity for each cryptid (incidentally, alternative identities noted in the addendum have not been included in the tables, as these refer to cryptids in Heuvelmans's list, not to those in this supplement), and also because there do not appear to be any "composite identity" examples present among the cryptids listed here.

As for the tables' taxonomic content: I have categorized cryptids down to the level of order within the phylum Chordata, and down to the level of class within all other (i.e. invertebrate) phyla, as I feel that to attempt anything more precise than this here would unavoidably introduce an unwieldy and unnecessary extent of taxonomic nomenclature in what is primarily a documentation of cryptids, not an exercise in defining or elucidating zoological taxonomy. For this same reason, I have chosen to ignore the wrath of taxonomy purists by adhering to familiar, traditional systems of classification—rather than complicating matters needlessly by introducing new (and often still controversial) versions that have yet to gain widespread recognition beyond their immediate circle of proponents and which would thus require extensive explanation in order to be readily comprehended here (a case in point is the radically revised classification system for birds inspired by molecular evidence).

TABLE 1:
Breakdown of Marine Cryptids in Supplement
Part 1: Phylum Chordata, divided into classes (C), orders (O), and suborders (So):

Incertae sedis: Class Mammalia?		1
C. Mammalia *Incertae sedis*: O. Pinnipedia?/O. Sirenia?		1
O. Cetacea:	So. Mysticeti	1
	So. Odontoceti	4
C. Osteichthyes *Incertae sedis*:		2
O. Stomiiformes		1
O. Gadiformes		1
O. Lophiiformes		1
O. Perciformes		1
O. Coelacanthiformes		1
C. Chondrichthyes O. Orectolobiformes		1
O. Isuriformes		1
O. Batiformes		1

Part 2: Invertebrate phyla (P), divided into classes (C):

P. Hemichordata		2
P. Chelicerata	C. Arachnida	1
P. Mollusca	C. Cephalopoda	1
P. Annelida	C. Polychaeta	1
P. Cnidaria	C. Scyphozoa	2
	TOTAL	**24**

TABLE 2: Breakdown of Freshwater Cryptids in Supplement

Phylum Chordata, divided into classes (C) and orders (O):

C. Mammalia *Incertae sedis*: O. Carnivora?/O. Sirenia?		1
	O. Carnivora	2
	O. Cetacea	1
C. Reptilia	O. Chelonia	1
C. Amphibia	O. Urodela	1
C. Osteichthyes *Incertae sedis*		2
	O. Salmoniformes	1
	O. Dipnoi	1
Incertae sedis: C. Osteichthyes?/C. Chondrichthyes?		1
	TOTAL	**11**

TABLE 3:
Breakdown of Terrestrial Cryptids in Supplement

Part A: Phylum Chordata, divided into classes (C), orders (O), and suborders (So):

Incertae sedis: Phylum Chordata?/Phylum Annelida?	1
C. Mammalia Incertae sedis: O. Carnivora?/O. Artiodactyla?	1
O. Marsupialia	4
O. Insectivora	1
O. Pholidota	1
O. Chiroptera	4
O. Primates	8
O. Carnivora	24
O. Perissodactyla	2
O. Artiodactyla	5
C. Aves *Incertae sedis*: O. Pelecaniformes?/O. Charadriiformes?; O. Ciconiiformes?/O. Galliformes?/O. Coraciiformes?; O. Falconiformes?/O. Gruiformes?/O. Strigiformes?/ O. Caprimulgiformes?; O. Apodiformes?/O. Passeriformes?	4
O. Phoenicopteriformes	1
O. Anseriformes	1
O. Falconiformes	1
O. Galliformes	5
O. Gruiformes	1
O. Psittaciformes	1
O. Musophagiformes	1
O. Strigiformes	1
O. Apodiformes	1
O. Passeriformes	9
C. Reptilia *Incertae sedis*: O. Squamata?/O. Sphenodontia?	1
O. Squamata: *Incertae sedis*: So. Sauria?/So. Serpentes?	1
So. Sauria	2
So. Serpentes	3
O. Pterosauria	3
C. Amphibia O. Salientia (Anura)	1

Part 2: Invertebrate Phyla (P), divided into classes (C):

P. Uniramia	C. Insecta	1
	TOTAL	**89**

Number of cryptids in supplement: 24 marine + 11 freshwater + 89 terrestrial = 124 total.

A LAST WORD

As demonstrated by this supplement, diligent bibliographical and field research can and will surely continue to yield reports of hitherto-unpublicized cryptids worthy of addition to Heuvelmans's original checklist—and, in turn, necessitate the preparation of further supplements. This research, therefore, should certainly be encouraged.

Having said that, however, I hope that publications such as the account presented here will ultimately lead not to an ever-increasing, but rather to an ever-decreasing, number of cryptids so listed — by actively stimulating investigations that climax in the formal discovery and scientific description of cryptids in Heuvelmans's list, in this current supplement, and in any future supplements. For this would finally bring to an end these animals' unjust, protracted exile from scientific respectability, and provide them with the coveted honor of official scientific recognition that they have so greatly, and for so very long, deserved.

As for those arch-cryptozoological skeptics who routinely dismiss reports of mystery beasts as mere fable and fantasy, yet who nevertheless will no doubt be reading this, I can but offer them the following valediction:

> *But remember always...that this is all a fairy-tale, and only fun and pretence; and, therefore, you are not to believe a word of it, even if it is true.*
> CHARLES KINGSLEY—THE WATER-BABIES

REFERENCES

ANON. (1968). New mammal discovered. *Animals*, 10 (March): 501-503.

ANON. (1976). Oliver...who? *Fortean Times*, No. 17 (August): 11, 20.

ANON. (1996). Not a potto. *Scientific American*, (April): 18.

ANON. (1997a). A Japanese wolf has survived!? *Sankei Shimbun* (evening edit.), 7 January.

ANON. (1997b). Oo-be-doo, I want to be like you [re Oliver]. *Fortean Times*, No. 95 (February): 15.

ANON. (1997c). Just a chimp off the old block. *Fortean Times*, No. 99 (June): 12.

ANON. (1997d). New monkeys found in Amazon. Associated Press report, 28 December.

ANON. (1998). Explorer's Tibetan trip in search of a duck. *Herald* (Glasgow), 2 June.

ALI, Salim & RIPLEY, S. Dillon (1969). *Handbook of the Birds of India and Pakistan, Together With Those of Nepal, Sikkim, Bhutan and Ceylon, Vol. 3* (2nd edit.). Oxford University Press (Bombay).

AMMANN, Karl (1993). Close encounters of the furred kind. *BBC Wildlife*, 11 (July): 14-15.

ANGIER, Natalie (1992). It may be elusive but moth with 15-inch tongue should be out there. *New York Times* (New York), 14 January.

ANNANDALE, N. (1908). An unknown lemur from the Lushai Hills, Assam. *Proceedings of the Zoological Society of London*, (17 November): 888-889.

BAUER, Aaron M. & RUSSELL, Anthony P. (1987). *Hoplodactylus delcourti* (Reptilia: Gekkonidae) and the *kawekaweau* of Maori folklore. *Journal of Ethnobiology*, 7 (summer): 83-91.

BEARDER, Simon (1997). Personal communications, 26 March, 1 April.

BEEBE, William (1924). *Galapagos: World's End*. G.P. Putnam's Sons (London).

BEEBE, William (1934). *Half-Mile Down*. Harcourt, Brace (New York).

BEEHLER, Bruce M. (1991). *A Naturalist In New Guinea*. Texas University Press (Texas).

BELDERSON, Martin (1995). Personal communication. 9 March.

BONET, Carlos (1995). Le *mamba mutu*: un carnivore aquatique dans le lac Tanganyika? *Cryptozoologia*, No. 10 (January): 1-5.

BONETTI, Andrea (1998). New life from Roman relics. *BBC Wildlife*, 16 (July): 10-16.

BROCK, Stanley E. (1963). *Hunting in the Wilderness.* Robert Hale (London).

CARAS, Roger (1975). *Dangerous To Man* (rev. edit.). Holt, Reinhart & Winston (New York).

CHORVINSKY, Mark (1995). Creature on ice. *Strange Magazine*, No. 15 (spring): 17.

CLARK, Jerome & COLEMAN, Loren (1978). *Creatures of the Outer Edge*. Warner (New York).

COLEMAN, Loren (1989). *Tom Slick and the Search For the Yeti*. Faber & Faber (London).

COURT, G.T. (1962). "Kondlo" - (a wild fowl). *African Wild Life*, 16 (December): 342.

DAVISON, G.W.H. (1983). Notes on the extinct *Argusianus bipunctatus* (Wood). *Bulletin - British Ornithologists Club*, 103(3): 86-88.

DAY, Sophy (1995). Personal communication. 22 May.

DOMENECH, Emanuel (1858). *Missionary Adventures in Texas and Mexico: A Personal Narrative of Six Years' Sojourn in Those Regions*. Longman, Brown, Green, Longmans, and Roberts (London).

DRATCH, P.A., *et al.* (1996). Molecular genetic identification of a Mexican onza specimen as a puma (*Puma concolor*). *Cryptozoology*, 12 [for 1993-1996, published in 1998]: 42-49.

ÉHIK, Gyula (1937-1938). Jackal or reed-wolf from Hungary. *Annales Musei Nationalis Hungarici, Pars Zoologica*, 31: 11-15.

ELLIS, Richard & McCOSKER, John E. (1991). *Great White Shark*. Stanford University Press (Stanford).

ELY, John J., *et al.* (1998). Technical report: chromosomal and mtDNA analysis of Oliver. *American Journal of Physical Anthropology*, 105(3): 395-403.

ELY, John J. & MOORE, Charleen M. (1998). Oliver update. *Science*, 280: 363-363.

FAWCETT, Percy H. (1953). *Exploration Fawcett*. Hutchinson (London).

FRICKE, Hans. (1989). Quastie im Baskenland? *Tauchen*, No. 10 (October): 64-67.

FULLER, Errol (1979). Hybridization amongst the Paradisaeidae. *Bulletin - British Ornithologists Club*, 99(4): 145-152.

FULLER, Errol (1995). *The Lost Birds of Paradise*. Swan Hill Press (Shrewsbury).

GIBBONS, Bill (1997). Personal communication. December.

GOSSE, Philip H. (1862). *The Romance of Natural History: Second Series*. Nisbet & Co (London).

GRAYSON, Mike (1998). The fortean fauna of Percy Fawcett. In: DOWNES, Jonathan & INGLIS, Graham (eds.), *The CFZ Yearbook 1998*. CFZ Publications (Exwick). pp. 23-33.

GREENWELL, J. Richard (1984). Interview [with Dr Bernard Heuvelmans]. *ISC Newsletter*, 3 (autumn): 1-6.

GREENWELL, J. Richard (1994a). The whatsit of Oz. *BBC Wildlife*, 12 (February): 53.

GREENWELL, J. Richard (1994b). She thought she saw a huia bird. *BBC Wildlife*, 12 (November): 26.

GREENWELL, J. Richard (ed.) (1986). Giant fish reported in China. *ISC Newsletter*, 5 (autumn): 7-8.

GREENWELL, J. Richard (ed.) (1987). Giant bear sought by Soviets. *ISC Newsletter*, 6 (winter): 6-7.

HALL, Mark A. (1992). Sobering sights of pink unknowns. *Wonders*, 1 (December): 60-64.

HEALY, Tony & CROPPER, Paul (1994). *Out of the Shadows: Mystery Animals of Australia*. Pan Macmillan Australia (Chippendale).

HELLABY, David (1984). Giant geckos 'sighted' 20 years ago. *Dominion* (Wellington), 11 September.

HEUVELMANS, Bernard (1955). *Sur la Piste des Bêtes Ignorées*. Plon (Paris).

HEUVELMANS, Bernard (1958). *On the Track of Unknown Animals*. Rupert Hart-Davis (London).

HEUVELMANS, Bernard (1968). *In the Wake of the Sea-Serpents*. Rupert Hart-Davis (London).

HEUVELMANS, Bernard (1982). What is cryptozoology? *Cryptozoology*, 1: 1-12.

HEUVELMANS, Bernard (1986). Annotated checklist of apparently unknown animals with which cryptozoology is concerned. *Cryptozoology*, 5: 1-26.

HEUVELMANS, Bernard (1996). Le bestiaire insolite de la cryptozoologie ou le catalogue de nos ignorances. *Criptozoologia*, No. 2: 3-18.

HILL, W.C. Osman (1953). *Primates. Comparative Anatomy and Taxonomy. I - Strepsirhini*. Edinburgh University Press (Edinburgh).

HOATH, Richard (1996). A deader desert. *BBC Wildlife*, 14 (September).

HOCKING, Peter J. (1992). Large Peruvian mammals unknown to zoology. *Cryptozoology*, 11: 38-50.

HOCKING, Peter J. (1996). Further investigations into unknown Peruvian mammals. *Cryptozoology*, 12 [for 1993-1996, published in 1998]: 50-57.

HOLDSWORTH, E.W.H. (1872). Note on a cetacean observed on the west coast of Ceylon. *Proceedings of the Zoological Society of London*, (15 April): 583-586.

IZZARD, Ralph (1951). *The Hunt For the Buru*. Hodder & Stoughton (London).

JACOBSON, E. (1937). The alovot, a bird probably living in the island of Simalur (Sumatra). *Temminckia*, 2: 159-160.

JANIS, Christine (1987). A reevaluation of some cryptozoological animals. *Cryptozoology*, 6: 115-118.

KINGDON, Jonathan (1990). Personal communication. 11 June.

LEMCHE, Henning, *et al.* (1976). Hadal life as analyzed from photographs. *Videnskabelige Meddelelser Dansk Naturhistorisk Forening*, 139: 263-336.

LEY, Willy (1941). *The Lungfish and the Unicorn: An Excursion Into Romantic Zoology.* Modern Age (New York).

LINDEN, Eugene (1994). Ancient creatures in a lost world. *Time*, (20 June): 56-57.

LYDEKKER, Richard (1899). On the supposed former existence of a sirenian in St. Helena. *Proceedings of the Zoological Society of London*, (20 June): 796-798.

MacFADDEN, B.J. (1980). Rafting mammals or drifting islands? Biogeography of the Greater Antillean insectivores *Nesophontes* and *Solenodon. Journal of Biogeography*, 7: 11-22.

MacINNES, Hamish (1979). *Look Behind the Ranges.* Hodder & Stoughton (London).

MACKAL, Roy P. (1980). *Searching For Hidden Animals.* Doubleday (Garden City).

MACKERLE, Ivan (1996). In search of the killer worm. *Fate*, 49 (June): 22-27.

MAGIN, Ulrich (1996). St George without a dragon. *Fortean Studies*, 3: 223-234.

MANGIACOPRA, Gary (1976). A monstrous jellyfish? *Of Sea and Shore*, 7 (autumn): 169.

MAREŠ, Jaroslav (1997). *Svet Tajemných Zvířat.* Littera Bohemica (Prague).

MARSCHALL, Count (1878-1879). Oiseau problématique. *Bulletin de la Société Philomatique* (7th Series), 3: 176.

MILLER, Gerrit S. (1930). Three small collections of mammals from Hispaniola. *Smithsonian Miscellaneous Collections*, 82: 1-10.

MUIRHEAD, Richard (1995). The flying snake of Namibia: an investigation. In: DOWNES, Jonathan (ed.), *CFZ Yearbook 1996.* CFZ (Exwick). pp. 112-123.

MURDOCH, W.G. Burn (1894). *From Edinburgh to the Antarctic.* Longmans, Green (London).

NOWAK, Ronald (1991). *Walker's Mammals of the World, Vol. I* (5th edit.). Johns Hopkins University Press (Baltimore).

OREN, David C. (1993). Did ground sloths survive in Recent times in the Amazon region? *Goeldiana Zoologia*, No. 19 (20 August): 1-11.

O'SHEA, Mark (1996). *A Guide to the Snakes of Papua New Guinea.* Independent Publishing (Port Moresby).

PFEIFFER, Pierre (1963). *Bivouacs à Borneo.* Flammarion (Paris).

PHILLIPPS, W.J. (1963). *The Book of the Huia.* Whitcomb & Tombs (Christchurch).

PITMAN, Robert L., *et al.* (1987). Observations of an unidentified beaked whale (*Mesoplodon* sp.) in the eastern tropical Pacific. *Marine Mammal Science*, 3 (October): 345-352.

PROROK, Byron de (1943). *Dead Men Do Tell Tales.* George Harrap (London).

RAKOTOARISON, Nasolo, *et al.* (1993). Lemurs in Bemaraha (World Heritage Landscape, Western Madagascar). *Oryx*, 27 (January): 35-40.

RAYNAL, Michel & MANGIACOPRA, Gary S. (1995). Out-of-place coelacanths. *Fortean Studies*, 2: 153-165.

RAYNAL, Michel & SYLVESTRE, Jean-Pierre (1991). Cetaceans with two dorsal fins. *Aquatic Mammals*, 17: 31-36.

RICHARDSON, Douglas (1995). Worldwide work. *Lifewatch*, (spring): 10.

RODGERS, Thomas L. (1962). Report of giant salamander in California. *Copeia*, 646-647.

ROESCH, Ben S. (1996). "The Thing": a cryptic polychaete of St. Lucia. *The Cryptozoology Review*, 1 (summer): 12-19.

RUTTER, Owen (1929). *The Pagans of North Borneo.* Hutchinson (London).

SANDERSON, Ivan T. (1973). More new cats? *Pursuit*, 6 (April): 35-36.

SANDERSON, Ivan T. (1974). The dire wolf. *Pursuit*, 7 (October): 91-94.

SCHROERING, Gerard B. (1995). Swamp deer resurfaces. *Wildlife Conservation*, (December): 22.

SCOTT, Peter (1983). *Travel Diaries of a Naturalist I.* Collins (London).

SEHM, Gunter G. (1996). On a possible unknown species of giant devil ray, *Manta* sp.. *Cryptozoology*, 12 [for 1993-1996, published in 1998]: 19-29.

SHUKER, Karl P.N. (1989). *Mystery Cats of the World.* Robert Hale (London).

SHUKER, Karl P.N. (1990a). A selection of mystery birds. *Avicultural Magazine*, 96 (spring): 30-40.

SHUKER, Karl P.N. (1990b). The Kellas cat: reviewing an enigma. *Cryptozoology*, 9: 26-40.

SHUKER, Karl P.N. (1991a). Gallinaceous mystery birds. *World Pheasant Association News*, No. 32 (May): 3-6.

SHUKER, Karl P.N. (1991b). *Extraordinary Animals Worldwide.* Robert Hale (London).

SHUKER, Karl P.N. (1993a). Hoofed mystery animals and other crypto-ungulates. Part III. *Strange Magazine*, No. 11 (spring-summer): 25-27, 48-50.

SHUKER, Karl P.N. (1993b). *The Lost Ark: New and Rediscovered Animals of the 20th Century.* HarperCollins (London).

SHUKER, Karl P.N. (1994). A belfry of crypto-bats. *Fortean Studies*, 1: 235-245.

SHUKER, Karl P.N. (1995a). Vietnam - why scientists are stunned. *Wild About Animals*, 7 (March): 32-33.

SHUKER, Karl P.N. (1995b). Menagerie of mystery. *Strange Magazine*, No. 16 (fall): 28-33, 48-49.

SHUKER, Karl P.N. (1995c). In the spotlight - the dobhar-chú. *Strange Magazine*, No. 16 (fall): 32-33, 49.

SHUKER, Karl P.N. (1995d). Snakes alive! It's the kting voar. *Wild About Animals*, 8 (October): 11.

SHUKER, Karl P.N. (1995e). Blue tigers, black tigers, and other Asian mystery cats. *Cat World*, No. 214 (December): 24-25.

SHUKER, Karl P.N. (1995f). *In Search of Prehistoric Survivors.* Blandford (London).

SHUKER, Karl P.N. (1996a). South American mystery cats. *Cat World*, No. 215 (January): 36-37.

SHUKER, Karl P.N. (1996b). A flight of fancy? *Wild About Animals*, 8 (January): 13.

SHUKER, Karl P.N. (1996c). Fins, fangs and poison. *Fortean Times*, No. 93 (December): 44.

SHUKER, Karl P.N. (1996d). Super-emus and spinifex men. *Fortean Studies*, 3: 189-210.

SHUKER, Karl P.N. (1997a). Potty about pottos. *Fortean Times*, No. 94 (January): 42.

SHUKER, Karl P.N. (1997b). Land of the lizard king. *Fortean Times*, No. 95 (February): 42-43.

SHUKER, Karl P.N. (1997c). The little howling god. *All About Dogs*, 2 (July-August): 46-47.

SHUKER, Karl P.N. (1997d). *From Flying Toads To Snakes With Wings.* Llewellyn Publications (St Paul).

SHUKER, Karl P.N. (1998a). Meet Mongolia's death worm: the shock of the new. *Fortean Studies*, 4 [the volume for 1997, but dated 1998]: 190-218.

SHUKER, Karl P.N. (1998b). Making the most of the migo [in Alien Zoo]. *Fortean Times*, No. 106 (January): 15.

SHUKER, Karl P.N. (1998c). Unmasking the onza — from Aztecs to Arizona. *All About Cats*, 5 (January-February): 44-45.

SHUKER, Karl P.N. (1998d). Shame about the shamanu [in Alien Zoo]. *Fortean Times*, No. 107 (February): 15.

SHUKER, Karl P.N. (1998e). A devil of a mystery [in Alien Zoo]. *Fortean Times*, No. 109 (April): 16.

SHUKER, Karl P.N. (1998f). A gravestone and an otter king. *History For All*, 1 (June-July): 48-49.

SHUKER, Karl P.N. (1998g). Turning turtle? [in Alien Zoo] *Fortean Times*, No. 113 (August): 18.

SHUKER, Karl P.N. (1998h). "A scaly tale from Rintja". *Fortean Times*, No. 116 (November): 45.

SHUKER, Karl P.N. (1999). *Mysteries of Planet Earth*. Carlton (London).

SHUKER, Karl P.N. (2000). Close encounters of the cryptozoological kind. *Fate*, 53 (May): 26-9.

SHUKER, Karl P.N. (2002a). Flying graverobbers. *Fortean Times*, No. 154 (January): 48-9.

SHUKER, Karl P.N. (2002b). *The New Zoo: New and Rediscovered Animals of the Twentieth Century*. House of Stratus (Thirsk).

SHUKER, Karl P.N. (2003). The deadly dodu [in Karl Shuker's Alien Zoo]. *Fortean Times*, No. 169 (April): 16.

SMITH, Malcolm (1996). *Bunyips and Bigfoots: In Search of Australia's Mystery Animals*. Millennium Books (Alexandria, NSW).

SOULE, Graham (1981). *Mystery Monsters of the Deep*. Franklin Watts (New York).

SPEIGHT, W.L. (1940). Mystery monsters in Africa. *Empire Review*, 71: 223-228.

SPENGEL, J.W. (1932). *Planctosphaera pelagica*. Scientific Results "Michael Sars" North Atlantic Deep-Sea Expedition, 5: 1-27.

TENNENT, J. Emerson (1861). *Sketches of the Natural History of Ceylon*. Longman, Green, Longman, and Roberts (London).

THIEL, Hjalmar & SHRIEVER, Gerd (1989). The DISCOL enigmatic species: a deep-sea pedipalp? *Senckenbergiana Maritima*, 20 (October): 171-175.

THOMAS, Lars (1996). No super-otter after all? *Fortean Studies*, 3: 234-236.

TISCHENDORF, Jay W. & ROPSKI, Steven J. (eds.) (1996). *Proceedings of the Eastern Cougar Conference, 1994*. American Ecological Research Institute (Fort Collins, Colorado).

TOHALL, Patrick (1948). The dobhar-chú tombstones of Glenade, Co. Leitrim. *Journal of the Royal Society of Antiquaries of Ireland*, 78: 127-129.

TRAJANO, E. & VIVO, M. de (1991). *Desmodus draculae* Morgan, Linares, and Ray, 1988, reported for Southeastern Brasil [sic], with pale-oecological comments (Phyllostomidae, Desmodontinae). *Mammalia*, 55(3): 456-459.

TRATZ, Eduard-Paul (1958). Ein Beitrag zum Kapitel "Rohrwolf", *Canis lupus minor* Mojsisovics, 1887. *Säugetierkundliche Mitteilungen*, 6: 160-162.

TROUGHTON, Ellis (1965). *Furred Animals of Australia* (8th edit.). Angus & Robertson (London).

WADE, Jeremy (1996). Tales from the bush. *BBC Wildlife*, 14 (June): 106.

WALLIS, Richard (1987). Letter to the editor. *British Herpetological Society Bulletin*, (autumn/winter): 65.

WALTERS, Michael (1980). *The Complete Birds of the World*. David & Charles (Newton Abbot).

WHITTY, Angus (director) (1995). *In Search of the Giant Flying Snake of Namibia* [TV documentary]. NNTV (Johannesburg).

WILLIAMS, John G. (1967). An unsolved mystery. *Animals*, 10 (June): 73-75.

WILLIAMS, John G. & ARLOTT, Norman (1980). *A Field Guide to the Birds of East Africa*. Collins (London).

WOOD, Casey A. & WETMORE, Alexander (1926). A collection of birds from the Fiji Islands. Part III: Field observations. *Ibis*, 2: 91-136.

YOUNG, Richard E. (1991). Chiroteuthid and related paralarvae from Hawaiian waters. *Bulletin of Marine Science*, 49 (1-2): 162-185.

ACKNOWLEDGMENTS

I am truly grateful to the following persons, publications, and organizations for their much-appreciated assistance and interest shown to me throughout my preparation of this book and the original articles upon which it is based.

All About Cats, All About Dogs, Animals and Men, Endymion Beer, Trevor Beer, Janet and Colin Bord, Mark Chorvinsky, Prof. John L. Cloudsley-Thompson, Loren Coleman, Gary Cunningham, Jonathan Downes, Drayton Manor Park & Zoo, *Enigmas*, Angel Morant Forés, *Fortean Studies, Fortean Times*, Philippa Foster, Richard Freeman, Bill Gibbons, Geoffrey R. Greed, J. Richard Greenwell, Paul Harris, Joe Harte, Bob Hay, David Heppell, Isabela Herranz, the late Dr. Bernard Heuvelmans, the late John Edwards Hill, *History For All*, Dr. Shoichi Hollie, Tyson Hughes, International Society of Cryptozoology, Dr. Ron Johnstone, Gerard van Leusden, Dr. John A. Long, Dr. Roy P. Mackal, Ivan Mackerle, Dr. Ralph Molnar, Steve Moore, Richard Muirhead, Darren Naish, Carina Norris, Heather Norris, Mark O'Shea, Harry Osoff, Michael Playfair, Alan Pringle, Marko Puljic, Dr. Clayton E. Ray, Michel Raynal, William M. Rebsamen, Bob Rickard, Ron J. Scarlett, Mary D. Shuker, Paul Sieveking, Malcolm Smith, *Strange Magazine*, Lars Thomas, Daev Walsh, *Wild About Animals*, Jan Williams, John G. Williams, *The X Factor*, and the Zoological Society of London.

In addition, I owe a special debt of gratitude, for their enduring support and encouragement, to my publisher Paraview, to my editor Patrick Huyghe, and to my agent, Mandy Little of Watson, Little Ltd.

SELECT BIBLIOGRAPHY

ANON. (1942). Riddle of the Orkney monster. *Orkney Blast* (Kirkwall), 30 January, pp. 1, 8.

ANON. (1942). Basking shark or "Scapasaurus"? *Orkney Blast* (Kirkwall), 6 February, pp. 1-2, 5.

ANON. (1962). Giant egg sets a problem for science. *The West Australian* (Perth), 3 May, p. 1.

ANON. (1970). The spinifex prints. *INFO Journal*, No. 7 (fall): 31-2.

ANON. [SANDERSON, Ivan T.] (1971). The abominable spinifex man. *Pursuit*, 4 (January): 9-10.

ANON. (1978). Moa hunters on visit. *Christchurch Press* (Christchurch), 16 February.

ANON. (1990). Skeleton of giant eagle discovered. *Daily Telegraph* (London), 17 January.

ANON. (1990). Une légende d'envergure. *Terre Sauvage*, No. 43 (September): 12.

ANON. (1991). The Scott River egg. Information sheet, Western Australian Museum (Perth).

ANON. (1994). Search [for] the Japanese wolf! *Nihon Keizai Shimbun* (Japan), 15 February.

ANON. (1994). The Japanese wolf surely survives! *Nihon Keizai Shimbun* (Japan), 20 March.

ANON. (1994). The beithir...folklore, cryptozoology...or what? *Athene*, No. 4 (September): 9.

ANON. (1994). More thoughts on beithirs and such. *Athene*, No. 5 (winter): 12-13.

ANON. (1996). [Rediscovery of Chinese ink monkey.] *People's Daily* (Beijing), 22 April.

ANON. (1996). Dragon, ahoy! *Fortean Times*, No. 89 (August): 13.

ANON. (1997). A Japanese wolf has survived!? *Sankei Shimbun* (evening edit.) (Japan), 7 January.

ANON. (1998). [Article re Greatstone sea serpent]. *Kentish Express* (London), 19 April.

ANON. (2001). Japanese 'rare wolf photo' was of dog, sign claims. Ananova News Service, 18 March.

ANON. (2002). 'Unicorn' bones in unmarked shipment. *Dominion Post* (New Zealand), 11 December.

ADAMS, W.H. Davenport (1885). *The Bird World*. Thomas Nelson (London).

ANDERSON, Atholl (1989). On evidence for the survival of moa in European times. *New Zealand Journal of Ecology*, 12 (supplement): 39-44.

ANDREWS, Roy Chapman (1926). *On the Trail of Ancient Man*. G.P. Putnam's Sons (London).

ASHTON, John (1890). *Curious Creatures in Zoology*. John C. Nimmo (London).

BALOUET, Jean-Christophe (1990). *Extinct Species of the World*. Charles Letts/David Bateman (London/Glenfield).

BARBER, Richard & RICHES, Anne (1971). *A Dictionary of Fabulous Beasts*. Macmillan (London).

BARCLAY, John (1811). Remarks on some parts of the animal that was cast ashore on the island of Stronsa, September 1808. *Memoirs of the Wernerian Natural History Society*, 1: 418-30.

BARLOY, Jean-Jacques & CIVET, Pierre (1980). *Fabuleux Oiseaux*. Robert Laffont (Paris).

BARNES, John (1817). *A Tour Through the Island of St. Helena*. J.M. Richardson (London).

BARTELS, Ernst & SANDERSON, Ivan T. (1966). The one true bat-man. *Fate*, 19 (July): 83-92.

BASKIN, Jon A. (1995). The giant flightless bird *Titanis walleri* (Aves: Phorusrhacidae) from the Pleistocene coastal plain of South Texas. *Journal of Vertebrate Paleontology*, 15: 842-4.

BAYANOV, Dmitri (1987). Why cryptozoology? *Cryptozoology*, 6: 1-7.

BEER, Rüdiger, R. (1977). *Unicorn: Myth and Reality*. James J. Kery (London).

BEISNER, Monika & LURIE, Alison (1981). *Fabulous Beasts*. Jonathan Cape (London).

BELLAIRS, Angus (1969). *The Life of Reptiles* (2 vols). Weidenfeld & Nicolson (London).

BENEDICT, Steve (1950). Man-eating trees. *Fate*, 3 (December): 13-16.

BENEDICT, W. Ritchie (1995). [Letter re Madagascan man-eating tree]. *Strange Magazine*, No. 15 (spring): 4.

BENNETT, Courtenay (1934). Monster or eel? [re Reunion giant eels] *The Field*, 163 (10 February): 289.

BERRY, Adrian (1989). Terrible Titanis bridges a gap in Ice Age evolution. *Daily Telegraph* (London), 13 February.

BEWICK, Thomas (1807). *A General History of Quadrupeds* (5th Edit.). E. Walker (London).

BLAND, K.P. & SWINNEY, G.N. (1978). Basking shark: genera *Halsydrus* Neill and *Scapasaurus* Marwick as synonyms of *Cetorhinus* Blainville. *Journal of Natural History*, 12: 133-5.

BORGES, Jorge L. & GUERRERO, Margarita (1974). *The Book of Imaginary Beings* (revised edit.). Penguin (Harmondsworth).

BREWSTER, Barney (1987). *Te Moa: The Life and Death of New Zealand's Unique Bird.* Nikau Press (Nelson).

BRIGGS, Katharine (1959). *The Anatomy of Puck.* Routledge & Kegan Paul (London).

BRIGHT, Michael (1989). *There are Giants in the Sea.* Robson (London).

BROOK, Stephen (2000). Rainbow serpent myth comes to life. *Australian,* 27 January, p. 1.

BROOKESMITH, Peter (Ed.) (1984). *Creatures From Elsewhere.* Orbis (London).

BUCKLAND, Frank T. (1858). *Curiosities of Natural History* (3rd Edit.). Richard Bentley (London).

BUDKER, Paul (1939). Sur la pretendue existence des phoques dans la région de l'Ile Shadwan (Mer Rouge). *Bulletin du Museum National d'Histoire Naturelle* (Séries 2), 11 (June): 450-3.

BURTON, Maurice (1959). *More Animal Legends.* Frederick Miller (London).

BUTLER, Harry (1969). Australia's embarrassing egg. *Science Digest,* 65 (March): 70-3.

BYRNE, M. St. Clare (Ed.) (1926). *The Elizabethan Zoo.* Frederick Etchells & Hugh MacDonald (London).

CLOUDSLEY-THOMPSON, John L. (1991). *Ecophysiology of Desert Arthropods and Reptiles.* Springer-Verlag (Berlin).

CLOUDSLEY-THOMPSON, John L. (1996). Reptiles that jump and fly. *British Herpetological Society Bulletin,* No. 55: 35-9.

COSTELLO, Peter (1974). *In Search of Lake Monsters.* Garnstone Press (London).

COSTELLO, Peter (1979). *The Magic Zoo: The Natural History of Fabulous Animals.* Sphere (London).

CUNNINGHAM, Gary (2003). The legend of the dobhar-chú. *Fortean Times,* No. 168 (March): 40-5.

DALE-GREEN, Patricia (1966). *Dog.* Rupert Hart-Davis (London).

DANCE, S. Peter (1976). *Animal Fakes and Frauds.* Sampson Low (Maidenhead).

DANQUAH, J.B. (1939). Living monster or fabulous animal? *West African Review,* (September): 19-20.

DAY, David (1981). *The Doomsday Book of Animals.* Ebury Press (London).

DERBYSHIRE, David (1998). Hen that crowed. *Daily Mail* (London), 4 March, p. 21.

DESFAYES, Michael (1987). Evidence for the ancient presence of the bald ibis *Geronticus eremita* in Greece. *Bulletin - British Ornithologists Club,* 107 (September): 93-4.

DETHIER, Michel & DETHIER-SAKAMOTO, Ayako (1987). The *tzuchinoko*, an unidentified snake from Japan. *Cryptozoology*, 6: 40-8.

DIAMOND, Antony W. *et al.* (1989). *Save the Birds* (rev. edit.). Pro Natur/ICBP (Frankfurt/Girton).

DZHAMBALDORZH, S. (1990). *Mongol Nuucyn Chelchee.* Ulan Bator.

EBERHART, George M. (2002). *Mysterious Creatures: A Guide to Cryptozoology* (2 vols). ABC-Clio (Santa Barbara).

EDWARDS, Alisdair (1990). *Fish and Fisheries of Saint Helena Island.* Centre for Tropical Coastal Management Studies, University of Newcastle upon Tyne (Newcastle).

EDWARDS, E.D. (1938). *The Dragon Book.* William Hodge (London).

EFREMOV, Ivan A. (1958). *Doroga Vetrov.* Editions Nationales des Manuels Pédagogiques (Moscow).

EMBODEN, William A. (1974). *Bizarre Plants - Magical, Monstrous, Mythical.* Studio Vista (London).

ESPINER, Colin (1993). Analysis shows moa to be deer - expert. *Christchurch Press* (Christchurch), 29 January, p. 1.

FAWCETT, Percy H. (1953). *Exploration Fawcett.* Hutchinson (London).

FLACOURT, Étienne de (1658). *Histoire de la Grande Isle Madagascar.* G. de Luyne (Paris).

FRASER, F.C. (1935). Zoological notes from the voyage of Peter Mundy, 1655-56. (b) Sea elephant on St. Helena. *Proceedings of the Linnean Society of London*, 147: 33-5.

FRASER, Pat (1991). Ongerup farmer's egg find sheds light on old mystery. *Great Southern Herald*, 19 June, p. 3.

FRIEND, Tim (1996). Fossil may be new link in primate evolutionary chain. *USA Today*, 5 April.

GILL, Brian (1991). *New Zealand's Extinct Birds.* Random Century New Zealand Ltd (Glenfield).

GRAHAM-DIXON, Andrew (1994). How much am I bid for a unicorn's horn? *Daily Mail* (London), 17 June, p. 9.

GREEN, Lawrence G. (1936). *Secret Africa.* Stanley Paul (London).

GREENHALL, Arthur M. & SCHMIDT, Uwe (Eds) (1988). *Natural History of Vampire Bats.* CRC Press (Boca Raton).

HADFIELD, Peter (2000). Is it a bird? Is it a plane? No one knows [re *tzuchinoko*]. *South China Morning Post* (China), 6 September.

HAND, Suzanne & ARCHER, Michael (Eds) (1987). *The Antipodean Ark.* Angus & Robertson (North Ryde).

HENRY, Walter (1843). *Events of a Military Life, Vol. 2* (2nd edit.). Pickering (London).

HEUVELMANS, Bernard (1958). *On the Track of Unknown Animals.* Rupert Hart-Davis (London).

HEUVELMANS, Bernard (1968). *In the Wake of the Sea-Serpents.* Rupert Hart-Davis (London).

HEUVELMANS, Bernard (1978). *Les Derniers Dragons d'Afrique.* Plon (Paris).

HIGHFIELD, Roger (1993). Australian scientists fail to unscramble mystery of the 2,000-year-old egg. *Daily Telegraph* (London), 20 July.

HOCKING, Peter J. (1992). Large Peruvian mammals unknown to zoology. *Cryptozoology*, 11: 38-50.

HUTTON, F.W. (1893). On Dinornis (?) queenslandiae. *Proceedings of the Linnaean Society of New South Wales*, 8: 7-10.

IDEMIZU, Nami (2000). Bounty put on mythical reptile. *Mainichi Shimbun* (Japan), 27 August.

IRWIN, J. O'Malley (1916). Is the Chinese dragon based on fact, not mythology? *Scientific American*, 114 (15 April): 399, 410.

KITCHING, G.C. (1936). The manatee of St. Helena. *Nature*, 138 (4 July): 33-4.

KOENIGSWALD, G.H. Ralph von & STEINBACHER, Joachim (1986). Fremde Vögel an Fernem Ort. *Natur und Museum*, 116 (April): 97-103.

KURTEN, Bjorn (1972). *The Ice Age.* Rupert Hart-Davis (London).

LEACH, John (1971). Man and wildlife in a Stone-Age culture. *Animals*, 13 (November): 868-71.

LEAKE, Jonathan (1997). Hunted: fabulous beasts of our time. *Sunday Times* (London), 23 March, p. 8.

LEY, Willy (1941). Scylla was a squid. *Natural History*, 48 (June): 11-13.

LODGE, G.A. (1976). King penguin egg washed ashore in Western Australia. *Western Australian Naturalist*, 13: 146.

LONG, John A. (1993). Eggstraordinary finds from Western Australia. *Australian Natural History*, 24 (summer): 6-7.

LONG, John L. (1981). *Introduced Birds of the World.* David & Charles (Newton Abbot).

LUM, Peter (1952). *Fabulous Beasts.* Thames & Hudson (London).

LYDEKKER, Richard (1899). On the supposed former existence of a sirenian in St. Helena. *Proceedings of the Zoological Society of London*, (20 June): 796-8.

MACKAL, Roy P. (1980). *Searching For Hidden Animals.* Doubleday (Garden City).

MACKAL, Roy P. (1987). *A Living Dinosaur? In Search of Mokele-Mbembe.* E.J. Brill (Leiden).

MACKERLE, Ivan (1994). In search of the killer worm of Mongolia. *World Explorer*, 1(4): 62-5.

MACKERLE, Ivan (1996). In search of the killer worm. *Fate*, 49 (June): 22-7.

MACKERLE, Ivan (1997). Cesta za alpským draken. *Fantastická Fakta*, (April): 6.

MACKERLE, Ivan (1997). Neznámé zvíře, nebo pohádkový netvor? [re Mackerle's *tatzelworm* expedition] *Fantastická Fakta*, (June): 6-7.

MACKERLE, Ivan (2001). The man-eating tree of Madagascar. *Fate*, 54 (July): 25-9.

MAGIN, Ulrich (1986). European dragons: the tatzelwurm. *Pursuit*, 19 (First Quarter): 16-22.

MAREŠ, Jaroslav (1993). *Legendární Příšery a Skutečná Zvířata.* Práce (Prague).

MARTIN, Geoffrey L. (1993). Elephant bird find nets family nest-egg. *Daily Telegraph* (London), 3 September.

MASEFIELD, John (ed.) (1906). *Dampier's Voyages, Vol. I.* E. Grant Richards (London).

MATTISON, Chris (1989). *Lizards of the World.* Blandford Press (London).

MAYOR, Adrienne (1991). Griffin bones: ancient folklore and paleontology. *Cryptozoology*, 10: 16-41.

MAYOR, Adrienne (2000). *The First Fossil Hunters: Paleontology in Greek and Roman Times.* Princeton University Press (Princeton).

MELLAND, Frank H. (1923). *In Witch-Bound Africa.* Seeley, Service & Co (London).

MELLISS, John C. (1875). *St. Helena: A Physical, Historical, and Topographical Description of the Island.* Reeve (London).

MICHELL, John & RICKARD, Robert J.M. (1982). *Living Wonders.* Thames & Hudson (London).

MILLER, Alden H. (1963). Fossil ratite birds of the Late Tertiary of South Africa. *Records of the South Australian Museum*, 14: 413-20.

MIRSKY, Jonathan (1996). Ink monkey of ancient China is rediscovered. *The Times* (London), 23 April.

MONCKTON, Charles A.W. (1922). *Last Days in New Guinea.* John Lane/The Bodley Head (London).

MORGAN, A.M. & SUTTON, J. (1928). A critical description of some recently discovered bones of the extinct Kangaroo Island emu (*Dromaius diemenianus*). *Emu*, 28: 1-19.

MORGAN, Gary, *et al.* (1988). New species of fossil vampire bats (Mammalia: Chiroptera: Desmodontidae) from Florida and Venezuela. *Proceedings of the Biological Society of Washington*, 101 (7 December): 912-28.

MORIARTY, Christopher (1978). *Eels: A Natural and Unnatural History.* David & Charles (Newton Abbot).

MORTENSEN, Theodor (1933). On the "manatee" of St. Helena. *Videnskabelige Middelelser fra Dansk Naturhistorisk Forening*, 97 (19 August): 1-9.

MORTENSEN, Theodor (1934). The 'manatee' of St. Helena. *Nature*, 133 (17 March): 417.

MURRAY, P.F. & MEGIRIAN, D. (1998). The skull of dromornithid birds: anatomical evidence for their relationship to Anseriformes (Dromornithidae, Anseriformes). *Records of the South Australian Museum*, 31: 51-97.

"N." (1851). Dragons. *Notes & Queries* (Series 1), 3 (18 January): 40-1.

NAAKTGEBOREN, C. (1990). Unicorns: immortal and extinct. *World Magazine*, No. 41 (September): 70-6.

NIGG, Joe (1982). *The Book of Gryphons*. Apple-wood Books (Cambridge, Massachusetts).

NIGG, Joe (1984). *A Guide to the Imaginary Birds of the World*. Apple-wood Books (Cambridge, Massachusetts).

NIGG, Joseph (Ed.) (1999). *The Book of Fabulous Beasts: A Treasury of Writings From Ancient Times to the Present*. Oxford University Press (Oxford).

O'FLAHERTY, Roderick (1846). *A Chorographical Description of West or H-Iar Connaught, Written A.D. 1684*. Irish Archaeological Society (Dublin).

O hÓGÁIN, Dáithí (1990). *Myth, Legend and Romance: An Encyclopaedia of the Irish Folk Tradition*. Ryan (London).

OLIVER, W.R.B. (1949). The moas of New Zealand and Australia. *Bulletin of the Dominion Museum, New Zealand*, 15: 1-206.

O'NEILL, June P. (1999). *The Great New England Sea Serpent*. Down East Books (Camden, Maine).

OSBORN, Chase S. (1924). *Madagascar, Land of the Man-Eating Tree*. Republic Publishing Company (New York).

PARKER, Shane A. (1984). The extinct Kangaroo Island emu, a hitherto unrecognized species. *Bulletin - British Ornithologists Club*, 104: 19-22.

PEACOCK, Edward (1884). The griffin. *The Antiquary*, 10 (September): 89-92.

PICKFORD, Martin & DAUPHIN, Yannicke (1993). *Diamantornis wardi*, nov. gen., nov. sp., giant extinct bird from Roilepel, Lower Miocene, Namibia. *Comptes Rendus de l'Académie des Sciences*, 316: 1643-50.

PROROK, Byron de (1934). 'Midst pygmies: exploring in unknown Mexico. *Wide World Magazine*, 73 (August).

PROROK, Byron de (1943). *Dead Men Do Tell Tales*. George Harrap (London).

RABINOWITZ, Alan (2001). *Beyond the Last Village: A Journey of Discovery in Asia's Forbidden Wilderness*. Island Press (Washington D.C.).

RAYNAL, Michel (1997). Olgoï-Khorkhoï, le 'ver-intestin' Mongol. *Cryptozoologia*, (June): 3-10, and *ibid.*, (July-August): 3-16.

REED, A.W. (1963). *Treasury of Maori Folklore*. A.H. & A.W. Reed (Wellington).

RICH, Pat V. (1979). The Dromornithidae, a family of large, extinct ground birds endemic to Australia. *Bureau of Mineral Resources Journal of Geology and Geophysics, Australasia*, 184: 1-196.

RICH, Pat V. (1981). Feathered leviathans. *Hemisphere*, 25 (May): 298-302.

RICH, Pat V. & TETS, Gerry F. van (Eds) (1985). *Kadimakara: Extinct Vertebrates of Australia*. Pioneer Design Studio (Lilydale).

"S., R." (1850). Dragons: their origins. *Notes & Queries* (Series 1), 2 (28 December): 517.

SAGAN, Carl (1977). *The Dragons of Eden*. Random House (New York).

SALVADORI, F. Baschieri & FLORIO, Pier L. (1978). *Wildlife in Peril*. Westbridge Books (Newton Abbot).

SANDERSON, Ivan T. (1937). *Animal Treasure*. Macmillan (London).

SANDERSON, Ivan T. (1972). *Investigating the Unexplained: A Compendium of Disquieting Mysteries of the Natural World*. Prentice-Hall (Eaglewood Cliffs).

SANDERSON, Ivan T. (1974). The dire wolf. *Pursuit*, 7 (October): 91-4.

SAUER, E.G. Franz & ROTHE, Peter (1972). Ratite eggshells from Lanzarote, Canary Islands. *Science*, 176 (7 April): 43-5.

SCARLETT, Ron J. (1969). On the alleged Queensland moa, *Dinornis queenslandiae* De Vis. *Memoirs of the Queensland Museum*, 15: 207-12.

SCHÖNHERR, Luis (1991). The tatzelwurm - mythical animal or reality? *Pursuit*, 22 (spring): 6-10.

SCHWARTZ, Randall (1974). *Carnivorous Plants*. Praeger Publishers (New York).

SCLATER, Peter (1980). *Rare and Vanishing Australian Birds*. Rigby (Sydney).

SEABURY, Charles (1852). Capture of the sea serpent. *The Times* (London), 10 March, p. 8.

SHARP, P.J. (1993). Orienting the unicorn. *BBC Wildlife*, 11 (October).

SHEPARD, Odell (1930). *The Lore of the Unicorn*. George Allen & Unwin (London).

SHOBERL, Frederic (Ed.) (1823). *The World in Miniature: Japan*. R. Ackermann (London).

SHUKER, Karl P.N. (1989). *Mystery Cats of the World*. Robert Hale (London).

SHUKER, Karl P.N. (1990). A selection of mystery birds. *Avicultural Magazine*, 96 (spring): 30-40.

SHUKER, Karl P.N. (1991). *Extraordinary Animals Worldwide*. Robert Hale (London).

SHUKER, Karl P.N. (1993). The case of the missing moa. *Fortean Times*, No. 69 (June-July): 42-3.

SHUKER, Karl P.N. (1993). Egg voyagers. *Fortean Times*, No. 70 (August-September): 44.

SHUKER, Karl P.N. (1995). Mongolian death worm - a shocking secret beneath the Gobi sands? *Strange Magazine*, No. 16 (fall): 31-2, 48.

SHUKER, Karl P.N. (1995). *Dragons: A Natural History*. Aurum Press (London).

SHUKER, Karl P.N. (1995). *In Search of Prehistoric Survivors*. Blandford Press (London).

SHUKER, Karl P.N. (1996). Highly charged. *Wild About Animals*, 8 (March): 9.

SHUKER, Karl P.N. (1996). *The Unexplained: An Illustrated Guide to the World's Natural and Paranormal Mysteries*. Carlton (London).

SHUKER, Karl P.N. (1996). Do nomads dream of electric worms? *In:* McNALLY, Joe & WALLIS, James (Eds). *Fortean Times Weird Year 1996*. John Brown Publishing (London): pp. 46-7.

SHUKER, Karl P.N. (1997). Otterly mysterious. *Wild About Animals*, 9 (January): 10.

SHUKER, Karl P.N. (1997). Mongolia's death worm: the shocking saga. *Uri Geller's Encounters*, No. 4 (February): 42-6.

SHUKER, Karl P.N. (1997). *From Flying Toads To Snakes With Wings*. Llewellyn (St Paul, Minnesota).

SHUKER, Karl P.N. (1998). A gravestone and an otter king. *History For All*, 1 (June-July): 48-9.

SHUKER, Karl P.N. (1999). *Mysteries of Planet Earth: An Encyclopedia of the Inexplicable*. Carlton (London).

SHUKER, Karl P.N. (2000). The skull of a wulver? *Fate*, 53 (November): 16-18.

SHUKER, Karl P.N. (2000). Death worms of the desert. *The X Factor*, No. 87: 2431-5.

SHUKER, Karl P.N. (2001). *The Hidden Powers of Animals*. Marshall Editions (London).

SHUKER, Karl P.N. (2002). *The New Zoo: New and Rediscovered Animals of the Twentieth Century*. House of Stratus (Thirsk).

SHUKER, Karl P.N. (Consultant) (1993). *Man and Beast*. Reader's Digest (Pleasantville).

SILLITOE, Paul (1981). Dance with the cassowaries. *Geographical Magazine*, 53 (May): 534, 537-8.

SKERRY, Brian (1997). High-seas drifter. *BBC Wildlife*, 15 (June): 64-5.

SKINNER, Charles M. (1911). *Myths and Legends of Flowers, Trees, Fruits, and Plants*. J.B. Lippincott (Philadelphia).

SPENCER, A. & KERSHAW, J.A. (1910). A collection of subfossil bird and marsupial remains from King Island, Bass Strait. *Memoirs of the National Museum, Melbourne*, 3: 5-35.

THOMPSON, Bill (1993). Roc of ages unscrambles egg mystery. *Sunday Times* (Perth), 21 March.

TINSLEY, Jim Bob (1987). *The Puma: Legendary Lion of the Americas*. Texas Western Press (El Paso).

TOHALL, Patrick (1948). The dobhar-chú tombstones of Glenade, Co. Leitrim. *Journal of the Royal Society of Antiquaries of Ireland*, 78: 127-9.

TRAJANO, E. & VIVO, M. (1991). *Desmodus draculae* Morgan, Linares, and Ray, 1988, reported for Southeastern Brazil, with paleoecological comments (Phyllostomidae, Desmodontinae). *Mammalia*, 55: 456-9.

TROTTI, Hugh H. (1987). Cryptoletter [re Marco Polo's hairy fish]. *ISC Newsletter*, 6 (summer): 11.

TSEVEGMID, Dondodijn (1987). *Altajn Tsaadakh Govd*. Ulan Bator.

TYSON, Peter (2000). *The Eighth Continent: Life, Death, and Discovery in the Lost World of Madagascar*. William Morrow (New York).

VAN DEUSEN, Hobart M. (1968). Carnivorous habits of *Hypsignathus monstrosus*. *Journal of Mammalogy*, 49: 335-6.

VERRILL, Hyatt & VERRILL, Ruth (1953). *America's Ancient Civilizations*. G.P. Putnam's Sons (New York).

VIS, Charles W. de (1884). The moa (Dinornis) in Australia. *Proceedings of the Royal Society of Queensland*, 1: 23-8.

WALKINGTON, L.A. (1896). A Bundoran legend. *Journal of the Royal Society of Antiquaries of Ireland*, 6: 84.

WAY, Mahalia (2003). The terrible griffin. *Fortean Times*, No. 170 (May): 50-5.

WELFARE, Simon & FAIRLEY, John (1980). *Arthur C. Clarke's Mysterious World*. Collins (London).

WENDT, Herbert (1959). *Out of Noah's Ark*. Weidenfeld & Nicolson (London).

WHITE, C.M.N. (1975). The problem of the cassowary in Ceram. *Bulletin - British Ornithologists Club*, 95: 165-70.

WHITE, C.M.N. (1976). The problem of the cassowary in New Britain. *Bulletin - British Ornithologists Club*, 96: 66-8.

WHITE, David G. (1991). *Myths of the Dog-Man*. University of Chicago Press (Chicago).

WHITLEY, Gilbert (1940). Mystery animals of Australia. *Australian Museum Magazine*, 7 (1 March): 132-9.

WILKINS, Harold T. (1952). *Secret Cities of Old South America*. Library Publishers (New York).

WILLIAMS, David L.G. (1981). *Genyornis* eggshell (Dromornithidae: Aves) from the Late Pleistocene of South Australia. *Alcheringa*, 5: 133-40.

WILLIAMS, John G. (1967). An unsolved mystery. *Animals*, 10 (June): 73-5.

WILSON, Andrew (1892). Science jottings. *Illustrated London News*, 50 (27 August): 279.

WILSON, Andrew (1892). Science jottings. *Illustrated London News*, 50 (24 September): 403.

WILSON, Dave (1993). Three claim sighting of moa. *Christchurch Press* (Christchurch), 25 January, p. 1.

WILSON, Dave (1993). Team on moa trail. *Christchurch Press* (Christchurch), 26 January, p. 1.

WILSON, Dave (1993). Computer fails to confirm moa. *Christchurch Press* (Christchurch), 12 February.

ZIMMER, Carl (1997). Terror take two [re *Titanis*]. *Discover*, (June): 68-74.

INDEX OF WILDLIFE NAMES

PARAVIEW publishes quality works that focus on body, mind, and spirit and the frontiers of science and culture. A complete list of our books and ordering information are available at www.paraview.com. Our cryptozoology titles include:

Bigfoot! The True Story of Apes in America
Loren Coleman $14.00/£8.99
Coleman takes readers on a journey into America's biggest mystery—could an unrecognized "ape" be living in our midst? He draws on more than 40 years of investigations, interviews, and fieldwork to reach some surprising conclusions about these animals—our nearest cousins!

Mysterious America: The Revised Edition
Loren Coleman $16.95/£12.99
Coleman has thoroughly updated his highly praised 1983 classic, which contains a never-before-published index; two completely new chapters; a new list; and updated and rewritten chapters. "Simply put, this is the best book I have ever read on the subject of America's mysteries."—Derek Anderson, *Stranger Things*.

Mothman and Other Curious Encounters
Loren Coleman $14.95/£11.99
On November 15, 1966, the appearance of "Mothman," a winged, red-eyed creature, set off 13 months of otherworldly mystery, madness, and mayhem. Coleman looks at its precursors, then brings the story up to date with recent sightings of its spawn.

In The Big Thicket: On the Trail of the Wild Man
Rob Riggs $13.95/£10.99
In the Big Thicket—the region of east Texas and southwestern Louisiana—ghost lights, phantom Indians, howling ape-like "wild men," and fireballs that streak through the nighttime skies defy both our commonsense notions of space and time and all attempts at scientific explanation.

PARAVIEW PUBLISHING offers books via three imprints. **PARAVIEW POCKET BOOKS** are traditionally published books co-published by Paraview and Simon & Schuster's Pocket Books. **PARAVIEW PRESS** and **PARAVIEW SPECIAL EDITIONS** use digital print-on-demand technology to create paperbacks for niche audiences. For more information, please visit our website at www.paraview.com, where you can also sign up for Conscious Planet, Paraview's free monthly media guide.

Printed in the United States
24510LVS00003B/52

9 781931 044646